Choosing Down Syndrome

Basic Bioethics
Arthur Caplan, editor

A complete list of the books in the Basic Bioethics series appears at the back of this book.

Choosing Down Syndrome

Ethics and New Prenatal Testing Technologies

Chris Kaposy

The MIT Press
Cambridge, Massachusetts
London, England

This book was set in Stone Serif by Westchester Publishing Services.

Library of Congress Cataloging-in-Publication Data

Names: Kaposy, Chris, author.
Title: Choosing Down syndrome : ethics and new prenatal testing technologies / Chris Kaposy.
Description: Cambridge, MA : The MIT Press, [2018] | Series: Basic bioethics | Includes bibliographical references and index.
Identifiers: LCCN 2017033895 | ISBN 9780262037716 (hardcover : alk. paper) ISBN 9780262546249 (paperback)
Subjects: LCSH: Down syndrome--Diagnosis--Moral and ethical aspects. | Prenatal diagnosis.
Classification: LCC RG629.D68 K37 2018 | DDC 616.85/8842--dc23 LC record available at https://lccn.loc.gov/2017033895

For Jan, Elizabeth, Aaron, and Ty

Contents

Preface

It is impossible to say how deeply we are indebted to those parents, children, teachers, and medical personnel who insisted on treating people with Down syndrome as if they *could* learn, as if they *could* lead "meaningful" lives. In bygone eras, the parents who didn't keep their children home didn't really have the "option" of doing so; you can't talk about options (in any substantial sense of the word) in an ideological current so strong. But in the early 1970s, some parents did swim upstream against all they were told and brought their children home, worked with them, held them, provided them physical therapy and "special learning" environments. These parents are saints and sages. They have, in the broadest sense of the phrase, uplifted the race. In the 10 million-year history of Down syndrome, they have allowed us to believe that we are finally getting somewhere.

—Michael Bérubé (1996, 27)

Here is where we are getting. Life expectancy for people with Down syndrome was less than ten years in the mid-1800s (Bérubé 1996, 27). In the 1960s, life expectancy for this group was around twenty years (Wright 2011, 185). Today, in resource-rich nations like the United States and Canada, people with Down syndrome live into their sixties on average (Wright 2011, 185).

In 1949, the estimated IQ of a child with Down syndrome living in an institution was 20–25 (Wright 2011, 95). Today close to 40 percent of people with Down syndrome have an IQ of 50–70, which classifies as a mild intellectual disability (Global Down Syndrome Foundation 2015). Many have an IQ in the normal range for the general population. Of course, life expectancy and IQ measures are largely social constructions. The increases that we see in life expectancy and IQ for people living with Down syndrome are attributable to greater acceptance of them into our communities and to the creation of social responses to their health and educational needs. The low life expectancy for people with Down syndrome prior to the 1970s is attributable to the effects of mass institutionalization, neglect within these institutions, and the refusal to treat infants with Down syndrome who had life-threatening though reversible illnesses at birth (Wright 2011).

I feel personally indebted to the saints and sages of the 1970s: the parents who—resisting pressure from doctors, family members, and friends—brought their children with Down syndrome home from the hospital and welcomed them into their families. In May 2009, my wife, Jan, gave birth to our son Aaron, who has Down syndrome. The saintly parents who have gone before us laid the groundwork for the joyous occasion of our son's birth. They made it easier not only to bring our child home from the hospital—today institutionalizing a child with Down syndrome is virtually unheard of—but also to accept and envision becoming a parent of a child with Down syndrome at an earlier stage of pregnancy, when Jan and I were given a prenatal diagnosis that Aaron had this condition. This news was difficult to hear and caused a fair amount of grief for both Jan and me. At the time, however, I was familiar with the history described by Bérubé. I was steeped in the philosophy of disability through my studies in bioethics. I had written a doctoral dissertation examining ethical issues involving infants with disabilities. I had heard the stories about parenting children with Down syndrome and read the data about the contributions that people with disabilities make to their families and their communities. These stories and this information made me believe it was possible not only to bring a child with Down syndrome into the world but also for our family to flourish with such a child—for Aaron's siblings to benefit from his presence in our lives.

At seven, Aaron is a happy little kid who enjoys playing ball hockey, watching YouTube videos, playing with his train set, and learning how to read. Aaron is especially devoted to his older sister, Elizabeth, and his younger brother, Ty. He looks forward to visits from his grandparents, and he misses them when they are not around. I put him to bed at night and get him ready for school in the morning. When I leave the house with him or walk him into school, Aaron dutifully holds my hand. At midday, Aaron eats lunch with his friends at school. To our surprise, he has learned his letters and the alphabet about as quickly as Elizabeth did, though he takes longer to learn other things, such as how to drink milk from an open cup without spilling it all over himself. In all of these respects, Aaron is very much like most other children who do not have a disability. He is good at some things and not so good at others. Our lives are better with him here with us. Our feelings toward Aaron are very similar to feelings that parents and siblings have to any child who does not have a disability. I know this because I feel the same way about Elizabeth and Ty.

This is a book about Down syndrome and prenatal testing. I present arguments for my position that more people should bring children with Down syndrome into the world, a position that has clearly been influenced by the presence of Aaron in my life and the life of our family. To some philosophers, my personal attachment to a child who has Down syndrome might seem to mean I cannot achieve the necessary "objectivity" about prenatal testing and Down syndrome for my analysis to be trusted. Licia Carlson, a philosopher who studies disability, has written about being questioned whether she has a family member with a disability, as though such a connection explains her research interests. Carlson states that, after being asked this question several times she "became increasingly irritated by the assumption that … the only reason I would have an interest in this topic is because someone in my family has an intellectual disability" (Carlson 2010, 2). The assumption implies that the lives of people with intellectual disabilities are not, in themselves, worthy of philosophical reflection—that such reflection could only be motivated by some personal relationship. Though Carlson does not in fact have a family member with such a disability, she suspects that her questioners also assume that "those who *do* have close personal ties to persons with intellectual disabilities are unable to achieve the distance required for objective and reasonable moral considerations" (Carlson 2010, 2).

My position is, first of all, that the lives of people with cognitive disabilities are of profound importance for philosophical study. Cognitive disability can cause a person to be vulnerable, and our responsibilities toward those who are vulnerable constitute a vitally important area of inquiry for ethics. The philosophical study of cognitive disability is crucial quite apart from any personal ties that individual philosophers may have with people with these disabilities. Second, as Carlson herself demonstrates, a family relationship with a person with a disability is not an admissions requirement for taking part in this discussion. Third, to discount the arguments of those who have close personal ties with people who have cognitive disabilities is to commit the ad hominem fallacy—one of the most elementary fallacies of critical thinking. I would hope that readers evaluate my arguments on their merits, not on the basis of judgments about my objectivity.

The arguments presented here are often about the experience of parenting children with Down syndrome. When it comes to discussions about what such experiences are like, parents of children with Down syndrome are actually in a privileged epistemic position. Rather than assuming that such parents are not objective, we should instead listen to them. Parents of children with Down syndrome know things about bringing children with this condition into the world and raising them that others do not know—or at least they have information to share to which others would have a different and difficult means of access. But even though I am such a parent, I do not regard my own experiences as a form of authoritative knowledge. The story of my own experiences only rises above the level of anecdote when it is supported by stories from multiple other sources, told by multiple other parents. As such, the voices of other parents of children with Down syndrome have a central place in my argument. These voices can be heard in the many parenting memoirs they have written, memoirs that are in turn supported, as I will show, by more formal social science research to which parents often contribute as key informants.

The issue of prenatal testing and "selective abortion"[1] for Down syndrome has become subject to legislation and extensive media attention recently in the United States. In 2014, the state of Pennsylvania adopted the Down Syndrome Prenatal Education Act, which requires that health care workers who counsel women receiving prenatal genetic testing must provide exclusively positive information about living with Down syndrome as well as information about support for raising a child with this condition (Caplan 2015). According to some commentators, this act signals a departure from the prevailing norms of ethical neutrality and "nondirectiveness" in prenatal genetic counseling and a move toward professional standards that are friendlier toward disabilities (e.g., Caplan 2015).

In another development, throughout the summer and fall of 2015, the state of Ohio considered adopting a law that would make it illegal to perform an abortion for the purposes of avoiding the birth of a child with Down syndrome (Lewin 2015). Although the proposed Ohio law failed to receive the votes it needed to pass, it is similar to one already adopted by North Dakota in 2013 (Lewin 2015). Opponents of these laws argue that they are simply a means for pro-lifers to set up new restrictions to abortions, and that they violate the U.S. Supreme Court's *Roe v. Wade* precedent establishing a woman's right to choose this procedure. Furthermore, they argue that a ban on selective abortion for Down syndrome is largely unenforceable (Lewin 2015). No one has been prosecuted under the North Dakota law (Lewin 2015). These developments made headlines in the U.S. media. The *New York Times* published an op-ed in September 2015 asking "Does Down Syndrome Justify Abortion?" (Schrad 2015).

This is not the first time Down syndrome has become the focus of bioethical debate. In fact, the abuse and neglect of people with Down syndrome has been a factor in some of the most significant cases in the development of bioethics as a subject of academic inquiry and professional regulation. In Bloomington, Indiana, in 1982, an infant was born who had Down syndrome complicated by an esophageal atresia, a discontinuity in the esophagus that prevents food from passing from the mouth to the stomach (Munson 2008). To the public, the child came to be known as "Baby Doe." Such an atresia can be corrected surgically, but because the infant had Down syndrome, the parents decided not to consent to treatment, which meant that the infant would die. The courts upheld this decision, and Baby Doe died of dehydration and other painful complications six days later (Munson 2008).[2]

The death of Baby Doe touched off a controversy. The U.S. federal government responded by issuing regulations requiring treatment for infants with disabilities who have life-threatening conditions (Robertson 2004). Hospitals and physician groups regarded this response as highly intrusive, and resisted these regulations—the "Baby Doe rules"—in the courts (Robertson 2004). The Baby Doe case was one of the historical causes for the creation of ethics committees in hospitals and of clinical ethics as a profession (Munson 2008). Although, these days, infants with Down syndrome who have treatable health problems are not left to die untreated, the influence of the Baby Doe case continues to be felt in discussions about whether to withhold or withdraw life-sustaining treatment for extremely premature infants specifically and in neonatal intensive care ethics generally (Robertson 2004; Lantos and Meadow 2006). Baby Doe was an "index case," a case that served as an early warning for changes that would eventually affect a great number of people. In this sense, the death of Baby Doe was a catalyst for the spread of formalized structures for clinical ethics.

The mistreatment of people with Down syndrome also figures in the history of research ethics. Willowbrook State School on Staten Island, New York, housed people with cognitive disabilities, including people with Down syndrome. Like many such institutions of the time, Willowbrook was understaffed, overcrowded, violent, unclean, and unsafe for its residents (Rothman and Rothman 1983). From the 1950s until the 1970s, a research team from New York University operated a program that involved injecting new residents with hepatitis viruses (Rothman 1982). Reasoning that, given the unhygienic nature of the institution, new residents to Willowbrook would have contracted hepatitis as a matter of course, the researchers considered it ethically acceptable to inject them directly with such viruses (Rothman 1982).

The horrors of Willowbrook were eventually exposed to the public, and its closure coincided with the deinstitutionalization movement in the treatment and care of people with cognitive disabilities like Down syndrome (Rothman and Rothman 1983). The saints and sages referred to by Bérubé played a major role in deinstitutionalization since they demonstrated that children with cognitive disabilities could be educated in the community and cared for at home. Institutionalization was all the more repugnant because the warehousing of people with cognitive disabilities was unnecessary. Along with the infamous Tuskegee syphilis experiment and other research ethics scandals of the time, the Willowbrook experiments were an impetus for the creation of research ethics review committees and of robust guidelines for the protection of vulnerable research subjects, such as those found in the *Belmont Report* (National Commission for the Protection of Human Subjects of Biomedical and Behavioral Research 1979). The discipline of bioethics has emerged, in part, because of the starvation, neglect, abuse, and deliberate infection of children with Down syndrome in health care and in medical research.

With the expansion of noninvasive prenatal testing (NIPT) as a tool for the identification of a whole panoply of fetal genetic differences, Down syndrome may now be serving as a new kind of index case. Down syndrome has been a prenatally diagnosable condition for decades. It was also among the first conditions that could be reliably identified with the use of NIPT, which involves a maternal blood draw without the risk of miscarriage. Innovation in noninvasive prenatal testing has enabled the creation and marketing of tests that identify rarer genetic conditions and genetic abnormalities arising from the addition or loss of genetic material that is much smaller than an "aneuploidy," or abnormal number of chromosomes (*PR Newswire* 2015). The genetic testing industry is quickly moving toward prenatal whole genome sequencing. Many new conditions have become detectable prenatally, such as Prader-Willi, Jacobsen's, and Klinefelter syndromes (Sequenom 2015). And genetic differences that correlate with a higher probability of autism will be identifiable prenatally very soon. As a result of innovation, people living with these conditions, their parents, and their advocates will be faced with the same ethical challenges and debates about the potential elimination of genetic differences from the human population that Down syndrome advocacy has been faced with for decades.

An index case alerts us to wider trends that begin to affect greater and greater numbers of people. The realization that some incident is not isolated, but rather presages developments that could ultimately affect any one of us or any of our loved ones can be a motivation for solidarity. I might feel moved to care about the other because the other appears to be just like me. The arguments I present in this book in favor of choosing to bring a child with Down syndrome into the world draw on two ideas that have solidarity at their heart. The first is the idea that those of us who do not have Down syndrome are linked with them in our shared vulnerability. Part of what it means to be human is to be dependent on one another—sometimes so highly dependent, such as in infancy and in illness, that we need care from others to keep us alive. The second idea is that people with Down syndrome should be seen not as "others" who can be dispensed with or removed from the human population by selective abortion. Because of our shared vulnerability in this world, people with Down syndrome are relevantly like us. Widespread selective abortion of fetuses with Down syndrome enables us to prevent the birth of a group of people perceived as different because of their level of dependency. In another sense, however selective abortion prevents the birth of people who are just like us *by virtue of* their dependency. Eliminating certain genotypes from the human population will not eliminate our dependency, even our extreme dependency, on one another.

Furthermore, the attempt to eliminate certain kinds of vulnerability and certain kinds of dependency—such as the dependency caused by Down syndrome—can also be harmful to those of us who do not have cognitive disabilities. The values that can lead us to see reproductive selectivity as a social good can be damaging when we ourselves are also vulnerable. I argue that values such as competitiveness, a focus on achievement, and economic opportunity are often in play when prospective parents make decisions about the kind of children they want to raise. These values—as well as biases against Down syndrome and cognitive disability generally—are apparent in some of the bioethical discussion about prenatal testing, selective abortion, and reproductive selectivity, and in the empirical research about reproductive decision making. Recognizing our human vulnerability, we should fear that values that can lead to the rejection of fetuses with vulnerable traits might also lead to the refusal of sympathy or assistance for our own plight.

To be sure, unsavory values are not the only possible motivations for prenatal testing or selective abortion. Some prospective parents faced with a prenatal diagnosis of Down syndrome have genuine concern for their other children, or are struggling financially, or are simply afraid and do not know much about Down syndrome. In the chapters to follow, I discuss all of these motivations and the social circumstances that give rise to them. My position is that we should be sympathetic to the needs and desires of prospective parents. There should be no legal limitations placed on access to selective termination or prenatal testing. At the same time, however, prospective parents should examine their own motivations carefully. It is likely that many selective termination decisions are based on misinformation about Down syndrome. The challenges of raising a child with Down syndrome are often overstated. I wrote *Choosing Down Syndrome* to improve information about Down syndrome available to the public.

My wife, Jan, and I decided to have Aaron for many different reasons. Jan felt a powerful bond with him while she was pregnant, and by the time he was prenatally diagnosed with Down syndrome, she already regarded him as our son. We also thought that we could handle the challenge. We viewed it to be an expectation of parenthood that we should accept our children for who they are, and not place conditions on our acceptance of them. We thought that if we were to be selective of our children, we would not be living up to what is required of parents. We saw a welcoming attitude toward our children as a virtue.

Aaron's birth was a celebration and an affirmation of our values. A neonatologist had warned us that infants with Down syndrome sometimes do not breathe on their own immediately after birth and sometimes require temporary assistance with breathing. But when Aaron was born, we were so excited because he screamed and breathed on his own without any problems. He has been a joy to parent every day since then.

It would be easy to say that choosing Aaron was a choice that we made, but that other people in the same situation might feel differently or have other values. As it stands, this position is fully justified. People should be free to make momentous decisions in line with what they value most. But, still, I would ask prospective parents: just what are your values? Have you examined them carefully? Maybe it is in fact consistent with your values to welcome a child with Down syndrome into your family.

Acknowledgments

I have many people to thank for the help they provided while I was writing this book. First of all, I have had several mentors, who have been influential in my development as a philosopher and bioethicist. Eva Feder Kittay, my doctoral supervisor, provided crucial advice at various stages. Jocelyn Downie, my postdoc supervisor, helped me develop good habits of argumentation. I hope these are to be found in this book. An important conversation with Susan Sherwin when I began writing steered me away from making some mistakes. Don Ihde was always ready to answer questions about publishing a book. Françoise Baylis included me in research meetings, where I could discuss my work, and asked me to write for her Impact Ethics blog, which provided a venue that I could use to develop some of my ideas.

My colleagues at Memorial University provided invaluable feedback and support. Jennifer Flynn, Fern Brunger, and Daryl Pullman, my fellow bioethics team members, and Associate Dean Shree Mulay have been the best colleagues anyone could ask for.

I presented portions and drafts of chapters from *Choosing Down Syndome* at the Canadian Philosophical Association meeting in 2010; the Canadian Bioethics Society conference in 2010 and 2015; the People's Health Matters lecture series at Memorial University in 2015; the Atlantic Regional Philosophical Association conference in 2016; the Caring for Philosophy conference in honor of Eva Feder Kittay in 2017; several Novel Tech Ethics research group meetings from 2010 to 2013; two Special Topics in Health Ethics classes hosted by Daryl Pullman and Jennifer Flynn, respectively, in 2015 and 2016; and Shree Mulay's Disease and Injury Prevention class in 2017. I had many excellent commentators during these interactions. Tim Krahn, Lynette Reid, Katie Wolfe, Diana Gustafson, and many Master of Health Ethics students at Memorial influenced how I shaped my arguments.

I would also like to thank Alice Crary for her encouragement, and Angel Petropanagos for her work on the Impact Ethics blog.

Memorial University supported my research with a sabbatical leave and a grant in the 2016–17 academic year. Philip Laughlin, who I always felt was on my side, shepherded the manuscript through the review process at the MIT Press. The thorough anonymous reviewers assigned by the Press judged my work critically but fairly; I revised the book several times in response to their insightful comments.

Finally, I owe my greatest debt of gratitude to my family: my wife, Jan, and my children, Elizabeth, Aaron, and Ty. They give meaning to everything I do.

A previous version of chapter 6 "Is Selective Termination Morally Wrong?," along with sections of chapters 1 and 2, were published as "A Disability Critique of the New Prenatal Test for Down Syndrome," copyright © 2013 The Johns Hopkins University Press, which first appeared in *Kennedy Institute of Ethics Journal* 23, no. 4 (December 2013): 299–324.

1 Introduction

People with Down syndrome have extra genetic material relating to the twenty-first chromosome. The most common form of Down syndrome is caused by the presence of a "trisomy," specifically, three copies of the twenty-first chromosome (trisomy 21), whereas the rest of the human population has two copies.[1] The addition of this extra genetic material happens randomly at the early stages of cell division after conception.

These genetic facts do not tell the whole story of Down syndrome. People with Down syndrome are able to develop, grow, learn, and live much like everyone else. They are beloved members of families. People with Down syndrome go to school, make friends, graduate, have jobs, get married, have sexual relationships, pursue hobbies and interests, start businesses, pursue higher education, and so on. It seems odd to create such a list since it is a list of common things that most people do. Such a list would be obvious and unnecessary when referring to people with a typical number of chromosomes. But because of their differences from others, people with Down syndrome, their supporters, and their organizations have to make a special effort to present the normality of their lives.

Down syndrome is associated with physical characteristics such as almond-shaped eyes, a relatively flat nose, short stature, and low muscle tone, though these characteristics are not shared by everyone who has the condition. The syndrome is also associated with a varying degree of cognitive disability—usually ranging from mild to moderate disability. Some children with Down syndrome have health problems, such as malformations of the heart and impairment of the thyroid, that are often attributed to the extra genetic material. But these health problems are also found among people who do not have the condition.

Down syndrome has been diagnosable prenatally since the late 1960s with the use of amniocentesis and karyotyping (Wright 2011), making it possible for prospective parents to decide whether to terminate pregnancies when the fetus has been shown to have the condition. *Choosing Down Syndrome* provides a line of reasoning for why prospective parents should instead decide to continue such pregnancies, bring children who have Down syndrome into the world, and welcome them into their families.

The case for choosing a child with Down syndrome can be summarized in fairly simple terms. The lives of people with Down syndrome tend to go well, just as they tend to do for people without this condition. Families that include people with Down syndrome tend to function as well as families that do not include a person with this condition. Raising a child with Down syndrome can fulfill parental expectations and be just as rewarding as raising a child without this condition. These reasons echo the efforts to establish in the public's mind the normality of the lives of people with Down syndrome, and this point about normality and difference recurs throughout this book. Though Down syndrome is a form of human difference, it is not a difference that in most cases would justify refusing to enter a parenting relationship with a child who has this condition.

Noninvasive Prenatal Testing

In the past few years, biotech companies have made noninvasive tests for Down syndrome available for clinical practice (Agarwal et al. 2013). These cell-free DNA-based tests require only a simple maternal blood sample as early as 10 weeks of gestation. In initial clinical trials involving thousands of pregnant women, Sequenom Inc.'s MaterniT21 test identified 99.1 percent of cases of Down syndrome (99.1% sensitivity), and gave the correct result in 99.9 percent of cases when the fetus did not have Down syndrome (99.9% specificity; Palomaki et al. 2011, 2012; Ehrich et al. 2011). Reports of high accuracy in these initial studies have since been tempered by more recent studies (Norton et al. 2015). One such study, published in the *New England Journal of Medicine*, found that the initial research cited above excluded subjects with inconclusive test results, and that, among these inconclusive tests, there was a higher likelihood that the pregnant woman was carrying a fetus with an aneuploidy such as Down syndrome (Norton et al. 2015). Much of the research into the accuracy of noninvasive prenatal testing (or NIPT[2])is either funded or conducted by the biotech firms that market the tests (Daley 2014). Because of the findings of the 2015 Norton study, the American College of Obstetricians and Gynecologists (ACOG) issued a statement about NIPT saying that "results [of NIPT] can be confusing as well as clinically uncertain" (ACOG 2015).

Nonetheless, the newer noninvasive tests are thought to be an improvement on previous prenatal testing techniques because of their accuracy early in pregnancy and because they don't carry a risk of miscarriage. Other testing techniques using maternal blood samples have only been able to give a probability that the pregnant woman is carrying a fetus with Down syndrome or other conditions. Invasive techniques that can give a definite diagnosis, such as amniocentesis and chorionic villus sampling (CVS), carry a small risk of miscarriage. At present, clinicians recommend that positive NIPT results should be confirmed by diagnostic tests like amniocentesis (ACOG 2015). In this way, NIPT can be used as a way of reducing the number of pregnant women who undergo amniocentesis, and are thus exposed to the risk of miscarriage. Some jurisdictions have incorporated, or plan to incorporate, NIPT into prenatal care as a second-tier screening test offered to pregnant women identified as having a high likelihood of carrying a fetus with Down syndrome or certain other conditions (Genetics Education Canada 2017; Perinatal Services BC 2017; United Kingdom Department of Health 2016). The positive predictive value of NIPT for Down syndrome is about 91 percent for this population, which is too low for the test to be regarded as diagnostic (United Kingdom National Screening Committee 2016).[3] Furthermore, the cost of NIPT is too high for it to be cost effective as a first-tier screening test (Morris et al. 2014). In the United Kingdom, the National Health Service plans to begin offering NIPT to pregnant women in 2018 if maternal serum screening[4] combined with ultrasound testing reveals a likelihood of 1 in 150 (0.7%) or greater of carrying a fetus with Down syndrome or two other trisomies (Nuffield Council on Bioethics 2017). Despite its relatively low positive predictive value even in "high likelihood" populations, many prospective parents are reported to regard NIPT as diagnostic (Buchanan et al. 2014; Daley 2014; Dar et al. 2014). In the future the positive predictive value of NIPT may be improved so that it becomes a diagnostic test.

In the United States, maternal blood samples are sent off to centralized laboratories for noninvasive prenatal testing. Patients can privately pay for the testing, which is also covered by some insurance plans (Agarwal 2013). NIPT is similarly available in Canada, the United Kingdom, and elsewhere to those willing to pay privately for it (Murdoch et al. 2017; Nuffield Council on Bioethics 2017). As mentioned, the United Kingdom is currently incorporating NIPT into its publicly funded National Health Service, and the Canadian provinces of British Columbia and Ontario have similar publicly funded health care services that offer NIPT as a second-tier screening test (Genetics Education Canada 2017; Perinatal Services BC 2017). Those who access NIPT through private payment or insurance coverage often use it as a first-tier screening test; when its findings are negative, they may then choose to forgo maternal serum screening.

The most up-to-date systematic review of research into selective termination rates within the U.S. population (Natoli et al. 2012) revealed that women terminated their pregnancies 60–90 percent of the time when prenatal tests diagnosed their fetus as having Down syndrome. The weighted mean termination rate for the studies included in this 2012 review was 67 percent, well below termination rates of around 90 percent after a prenatal diagnosis of Down syndrome documented in a systematic review of international studies thirteen years before (Mansfield, Hopfer, and Marteau 1999). In the United Kingdom, the selective termination rate after a diagnosis of Down syndrome has remained stable between 89 and 95 percent since 1989 (Nuffield Council on Bioethics 2017). As I will argue in chapter 2, there are features of the new generation of noninvasive tests that could contribute to an increase in the total numbers of selective terminations after a diagnosis of Down syndrome.

A Pro-Choice Position

Choosing Down Syndrome presents the case for bringing children with Down syndrome into the world. That said, my position with regard to abortion is strongly pro-choice. I am not advocating that access to selective abortion or to prenatal tests for detecting Down syndrome be restricted. I am also not arguing that it is ethically mandatory in all cases either to give birth to a child with Down syndrome when diagnosed prenatally or to forgo prenatal testing. I do not argue that selective termination is immoral in itself. There are some cases in which selective termination seems justifiable. I discuss this point in particular in chapter 7.

My goal is to show why having a child with Down syndrome would likely be a rewarding experience for most prospective parents. *Choosing Down Syndrome* can be read as a direct address to prospective parents and to those who come into contact with them (friends, family, health care providers) about why more people should have a child with Down syndrome. I use the tools of philosophical analysis to examine some of the possible reasons people find compelling as justifications for using prenatal testing and selective termination. I suggest that much of the reasoning supporting the choice to test and terminate is influenced by misinformation and faulty inferences or by unconscious political ideologies or biases that prospective parents might well disavow if they became aware of them.

Though my position is pro-choice, I do not believe that personal reproductive decisions should be beyond the reach of ethical reflection or examination. People should be free to make choices about their reproductive lives based on whatever reasons they have. But the freedom of choice should not prevent respectful rational engagement with the beliefs, motivations, and arguments that inform reproductive choices. Though reproductive choices are deeply personal, there is nothing ethically wrong with examining the reasoning that underlies these choices. I draw upon information about this reasoning from several sources. For instance, there is empirical research about actual selective abortion choices people have made (e.g., Korenromp et al. 2007). Furthermore, some research has explored the justifications people would give if asked to make a decision about a hypothetical scenario in which selective abortion is an option (Bryant, Green, and Hewison 2010; Choi, Van Ripper, and Thoyre 2012). Many people have also written personal testimonials and memoirs about prenatal decision making and raising a child with Down syndrome (see chapter 3). Contributors to the philosophical and bioethics literature on selective reproduction also provide rational reconstructions of decision making (e.g., Buchanan et al. 2000; Asch and Wasserman 2005; Wilkinson 2010).

A powerful example of reasoning about prenatal testing and selective abortion comes from Mary Ann Baily (2000), who describes why she had undergone prenatal testing in the past, and why she would have terminated her pregnancy if testing revealed that the fetus had a disability such as Down syndrome. Baily (2000, 68) says simply that, "given a choice, I would rather my child did not have a disability. That's all." Though the reason she gives might seem immediately compelling to some, it bears further analysis. For instance, one might wonder, with Deborah Kent, why disability in itself should be seen as "fundamentally undesirable" (Kent 2000, 58–59). Fortunately, Baily (2000) explains her simple reason and provides further insight into her thinking; she states that a child born with Down syndrome will have "potentially significant limitations" (66). She states as well that, even though families of children with Down syndrome tend to function well, "life will still be more difficult" for a family with such a child (70). In the chapters that follow, I will discuss each of these assertions.

In particular, I will assess what it means for a child to have "serious limitations." In the case of a child who has Down syndrome, the main limitation is cognitive disability. I will analyze the consequences of this limitation for the child's well-being and for the prospective parents' socially conditioned expectations about their children. In short, on the basis of empirical evidence, I will show that cognitive disability associated with Down syndrome cannot be expected to lead to diminished well-being. Furthermore, parental expectations about children that might support selective abortion are often based on unacknowledged beliefs about the worth of human beings—beliefs that deserve to be brought into the light of day so that prospective parents can decide whether they actually endorse them.

On the subject of family functioning, I show that the increased difficulty of caring for a child with Down syndrome is either overstated or usually regarded as insignificant by these families themselves. This analysis of reasoning within the private sphere of reproductive decision making is meant to contribute to the "social fund of knowledge about disability," which "widen[s] the space of possibility in which relationships can be imagined and resources claimed" (Ginsburg and Rapp 2010, 239). My goal is to increase the social fund of knowledge, and, in doing so, to show prospective parents that they can do it—it is possible and desirable to have a child with Down syndrome.

The Problem

The problem that needs to be addressed might not have become apparent thus far. What is the benefit to be achieved by increasing the births of people with Down syndrome? I am not defending the idea of fetal harm caused by abortion. I don't believe that fetuses have moral standing such that they can be harmed by terminating a pregnancy. The harm that needs to be addressed is caused by a common bias against people with Down syndrome and other cognitive disabilities that persists in our culture. This bias manifests in many ways, such as the prevalence of the terms "retarded" or "retard" in everyday language. Although many people have taken up the challenge of eliminating them from common speech, these slurs are still socially accepted to a greater degree than slurs directed at people of other races or religions or at women or gay people.

Though the quality of life of people with Down syndrome tends to be quite good, the bias against them has harmful consequences. As I show in chapter 6, empirical researchers have found an alarming degree of loneliness and friendlessness among people with Down syndrome and other cognitive disabilities. Referring to the social isolation of people with such disabilities, Marsha Saxton (2010, 122) states, "It is discriminatory attitudes and thoughtless behaviors, and the ensuing ostracism and lack of accommodation that make life difficult" for people with the disabilities, rather than the disabilities themselves. The creation of Down syndrome societies and other social support groups has been a vital social outlet for many people with this condition who might feel isolated when nondisabled friends from early childhood drift away, or who have few social links outside of their families. But the acceptance and inclusion of people with Down syndrome could be further improved by the elimination of bias.

People with Down syndrome also suffer when bias against them makes it harder to get a job or find suitable housing. As Adrienne Asch (2003, 327) points out, there are significant gaps "between people with and without disabilities in terms of education, employment, income, social life, and civic participation." There should be little wonder that these gaps exist since bias directed at people with Down syndrome and other cognitive disabilities can certainly influence decisions in all areas that affect the lives of people with these conditions. Furthermore, a large proportion of public medical research funding that could go toward promoting the health of people with Down syndrome is instead directed toward developing tests for identifying the condition prenatally (Krahn 2015). The common bias against people with Down syndrome is a discernible motive underlying these gaps and discriminatory policies.

Biased attitudes toward Down syndrome are not caused by the availability of prenatal testing and selective termination, or at least there is no intuitive causal link. Rather, it is the other way around: biased attitudes are themselves a cause of the uptake of prenatal testing and selective termination. Actions that arise out of biased motivations also perpetuate attitudes that are harmful to people with Down syndrome. The arguments in *Choosing Down Syndrome* are meant to counter these biased attitudes and the harm they cause. Reducing them would improve the lives of people who have Down syndrome and lead to more births of children with this condition. Though there may not be a causal link between the introduction of noninvasive prenatal testing and biased attitudes against people with disabilities, there is benefit to countering biased attitudes wherever they manifest. I argue (in chapter 6) that bias plays a role in decisions about whether to continue a pregnancy when the fetus has been diagnosed with Down syndrome—not in all cases, but inevitably in many such decisions given the prevalence of this bias in our culture.

If more people have children with Down syndrome because they refuse prenatal testing or selective termination, the greater number of people with this condition and of families caring for them could help address some of the problems I have identified. A greater number of people with Down syndrome would bring with it a greater number of ambassadors and advocates advancing their interests, reducing bias, supporting social inclusion, better housing, and better employment.

Other Disabilities

The arguments and conclusions I reach in this book, strictly speaking, apply only to Down syndrome. However, as I state in chapter 3, similar arguments could be formulated in support of bringing children with other disabilities into the world—disabilities that can now be detected prenatally or that will be in the future. Although I don't fully explore this possibility, a few observations are worth making here.

The reasoning I have developed about Down syndrome depends on empirical findings and arguments about family functioning and the well-being of people with Down syndrome. The extent to which a similar case could be made for people with other disabilities will depend on the support provided by social science research. Of course absent this support, other arguments could be made for the recommendation that prospective parents should bring a child with a given disability into the world.

On the issue of family functioning, families with children who have Down syndrome are often studied alongside families with children who have other disabilities—such as autism. Gillian A. King and colleagues (2006) report positive findings about the functioning of families that include children with Down syndrome as well as of families that include children with autism. However other research in this area has arrived at mixed conclusions. Some studies (e.g., Abbeduto et al. 2004) describe a "Down syndrome advantage," in which families that include a child with Down syndrome typically have better functioning than families with children who have other disabilities.

Many of the research findings on the well-being of people with disabilities apply not only to Down syndrome but also to many other disabilities. These findings tend to show that people with disabilities report good quality of life (e.g., Ubel et al. 2005; Albrecht and Devlieger 1999). There is also disability-specific literature showing that it is common for people with even severe disabilities, such as severe spina bifida, to experience well-being (Sawin et al. 2002; Padua et al. 2002).

A third crucial element of my argument as it pertains to Down syndrome is the stigma attached to people with cognitive disabilities in our culture. The harms that motivate my recommendation that more people should choose to have a child with Down syndrome derive from the pervasiveness of this stigma. Arguments similar to mine will be of limited applicability to other prenatally diagnosable conditions if these conditions are not also stigmatized. Cystic fibrosis might be an example (Robinson 2002), as well as late-onset conditions such as Huntington's disease or Gerstmann-Sträussler-Scheinker disease (Kolata 2014)—though people more knowledgeable about these conditions than I might argue that bias is often directed at people with such conditions as well.

Other Genetic Tests

Preimplantation Genetic Diagnosis

Just as the central arguments in this book are limited to Down syndrome, they are also limited to examining the ethics of prenatal testing technologies such as NIPT and of selective abortion. There are important differences, for instance, between noninvasive prenatal testing and preimplantation genetic diagnosis (PGD), which involves testing the genetic material of implantable embryos created through in vitro fertilization (IVF) before their implantation. I am reluctant to apply the arguments I have developed in this book to PGD scenarios. Consider a scenario in which one embryo that is a candidate for implantation has been determined to have trisomy 21, the genetic cause of the most common form of Down syndrome. My position is that implanting such an embryo is fully justified and understandable in light of the well-being of people with Down syndrome and their contributions to the well-being of their families and communities. Implanting an embryo with Down syndrome would also be a positive contribution (as I have argued) to efforts to remove the stigma attached to this condition. But this position runs contrary, for example, to that of the United Kingdom's Human Fertilisation and Embryology Act of 2008, section 14(4), which prohibits selection of an embryo that has the genetic traits that would cause a disabling condition.

Quite unlike a decision to selectively terminate a pregnancy, a decision *not* to implant an embryo determined to have Down syndrome, is not subject to many of the ethical concerns outlined in this book. In the preimplantation genetic diagnosis scenario, prospective parents would not be rejecting a previously wanted pregnancy simply on the basis of a diagnosis, as they would with NIPT and selective termination. With PGD, the woman is not pregnant prior to implantation. The refusal to implant an embryo with Down syndrome in a PGD scenario thus seems much less tied to biased attitudes toward people with cognitive disabilities because a refusal to implant is a much less extreme step than the termination of a previously wanted pregnancy.

Nonetheless, the decision to test embryos for Down syndrome prior to implantation might just as well be a product of bias. If such testing is or becomes a routine part of in vitro fertilization, similar concerns about bias toward Down syndrome might apply to PGD as well. I can state with confidence, however, that the decision to choose an embryo with Down syndrome would be justified and positive for the community of people with this condition. But a full defense for the position that pregnant women should not refuse to implant such an embryo would require arguments different from those presented in this book.

Expanded Noninvasive Prenatal Testing

The development of NIPT for Down syndrome is just one advance in a fast-moving area of the biotech industry. As mentioned in the preface, the NIPT portfolio among test developers has already increased to include tests for such conditions as trisomy 18 and Prader-Willi, Jacobsen's, and Klinefelter syndromes (Sequenom 2015). Many of these conditions were not previously detectable with prenatal testing. The company Sequenom has introduced their MaterniT Genome test, which is able to identify smaller gains or losses in genetic material "greater than 7 megabases in size" across the entire fetal genome, using only a sample from a maternal blood draw (*PR Newswire* 2015). The testing industry is moving toward prenatal whole-genome sequencing of the fetus, "allowing quick and accurate analysis of every base pair, as well as mitochondria" (Munthe 2015, 37). Prenatal whole-genome sequencing used in tandem with gene-editing technology like CRISPR (clustered regularly interspaced short palindromic repeats) and with in vitro fertilization could give prospective parents unprecedented control over the genomes of their prospective children (Ledford 2015). There are also efforts to improve the ease of testing. For example, a biotech company in the United Kingdom has reported initial success of a maternal urine test for detecting Down syndrome (Iles et al. 2015).

The increase in the scope of testing will mean that even more conditions will be detectable prenatally, though current tests for rare conditions are of doubtful accuracy (Munthe 2015). These developments raise a host of new ethical issues. Although I am unable to discuss the issues at any great length here, it is worth briefly considering several of them. Achieving informed consent, for example, becomes challenging for health care providers offering noninvasive prenatal testing to test for a wide range of conditions when the accuracy of the tests varies, when the data supporting the reported accuracy levels are uncertain, and when the conditions are of different degrees of severity. Some of these problems could arise if NIPT is developed as a means toward prenatal whole-genome sequencing (Donley, Hull, and Berkman 2012; Chen and Wasserman 2017). Informed consent will require extensive contact with highly trained genetic counselors, and improved data (Meredith et al. 2016). Many jurisdictions appear unprepared for these challenges due to insufficient numbers of genetic counseling professionals (Meredith et al. 2016; Bernhardt 2014; Carroll 2009). Informed consent may also become compromised by the routinization of NIPT. Procedures that are now standard aspects of prenatal care such as maternal serum screening and ultrasound already have weakened norms of informed consent such that many women undergo these procedures without truly knowing the risks and benefits (Seavilleklein 2009). Some worry that a similar routinization of NIPT could result in women being pressured to undergo this testing, or could result in a lack of awareness that the test is even being conducted (Murdoch et al. 2017; Kaposy 2013). Another concern is that the termination of pregnancies that test positive through NIPT could be cast as the "normal" or expected outcome of NIPT (Murdoch et al. 2017).

The implementation of noninvasive prenatal testing is also ethically controversial. The biotech companies profiting from these tests are pushing for the adoption of NIPT as a first-tier screening test that should be offered to all pregnant women (Karow 2016). This advocacy, along with the industry-funded nature of much of the research into NIPT, creates the possibility of conflicts of interest, and corporate-induced biases at the expense of good patient care and cost-effectiveness. The adoption of publicly funded programs for NIPT carries other ethical challenges, such as the difficult delineation of what constitutes a genetic condition serious enough to justify a publicly funded program for its prenatal detection. Should such a program be offered as a way of advancing reproductive autonomy or as a public health initiative? (I examine this question briefly in chapter 2). Overall, the lack of substantial regulatory oversight of these new tests (also discussed in chapter 2) is an ethical concern when health care systems attempt to incorporate NIPT into prenatal care, either with or without public funding.

The prospect of probabilistic diagnostic information for conditions like autism is especially frightening. Consider a scenario in which NIPT used in conjunction with whole-genome sequencing can accurately reveal that a fetus has a cluster of genes that correlates with a numerical probability for the development of autism. This scenario is different from one in which tests give probabilistic information because they aren't accurate or don't have high positive predictive value. Selective termination decisions will become highly fraught, stressful, and contentious if tests are developed and made available to the public that reveal accurate but only probabilistic information. New generations of tests will complicate reproductive decisions by the expansion of information about the fetus that will be available prenatally.

In various places throughout the chapters that follow, I address some of these ethical issues related to noninvasive prenatal testing: informed consent, implementation, regulation, corporate involvement, and public funding. Nonetheless, I do not attempt anything like a comprehensive analysis of these issues, nor do I intend this book to be a comprehensive text on ethical challenges related to NIPT. My focus, rather, is on Down syndrome in particular, and on the ethical arguments supporting my position that more prospective parents should choose to have children with Down syndrome, even though NIPT and other technologies enable them to choose otherwise.

The Chapters

Chapter 2 situates the development of noninvasive prenatal testing within the recent history of ethical debates about prenatal testing and selective abortion. It articulates the terms of these debates, my position with respect to major issues, and introduces those unfamiliar with these precedents into the ongoing discussion as it has played out prior to the invention of NIPT. I discuss the "expressivist" objection to prenatal testing and selective abortion, and address analogical arguments that liken selective abortion to sex selection or to folic acid supplementation. I introduce and defend my position on the impact of cognitive disability on a person's well-being, which has been a contentious issue in reproductive ethics. The chapter ends with a discussion of the ethics of implementing NIPT, which has been the subject of current debate.

Drawing on memoirs and essays written by parents of children with Down syndrome as well as on social science sources, chapter 3 describes the experience of parenting a child with Down syndrome through the first days and years of life. The lively and growing genre of Down syndrome parent memoirs provides detailed and intimate accounts of living with a child who has Down syndrome. A common theme emerges from the seven book-length memoirs and ninety-six autobiographical essays studied for this chapter. Parents who have children with Down syndrome commonly describe how the birth of their children transformed them and altered their value systems in profound ways. The authors of these works universally endorse these changes. They see themselves and their families as having benefited from the addition of children who have Down syndrome.

Chapter 4 deals with a possible response to the positive assessments of parenting a child with Down syndrome in the parenting memoirs and essays presented in chapter 3. This possible response is to allege that the positive assessments are marked by self-deception or adaptive preference that distorts their evaluation of family functioning. This chapter therefore counters the allegation that Down syndrome parenting memoirs are unreliable indicators of the typical functioning of such families. To counter this allegation, the chapter presents social science findings that support and largely reinforce the evidence presented in the Down syndrome parenting memoirs and essays. Although some of the social science findings are based on subjective assessments of family functioning by the parents themselves, many are based on more objective indicators such as studies of divorce rates, observational studies, and analyses of the behavior of siblings of children with Down syndrome. The social science findings are almost uniformly positive. Families of children with Down syndrome typically function as well as families with only nondisabled children.

Chapter 5 discusses arguments that it is morally wrong to bring a child with Down syndrome into the world when prospective parents could try again for a nondisabled child instead. One such argument, presented by Allen Buchanan, Dan W. Brock, Norman Daniels, and Daniel Wikler in their book *From Chance to Choice*, alleges that prospective parents cause "wrongful disability" when they choose to continue a pregnancy with a fetus diagnosed as having Down syndrome. Julian Savulescu and Guy Kahane (2009) make a similar argument in their defense of the "Principle of Procreative Beneficence." The chapter counters the "wrongful disability" argument in *From Chance to Choice* by showing that there is little reason to accept its views about the well-being of people with disabilities and that its prioritization of economic opportunity above all other values and social contributions made by people with disabilities is unjustifiable. I argue that Savulescu and Kahane's account is unconvincing because of the questionable assumptions they make about the nature of human well-being.

Following my discussion of the argument that it is morally wrong not to selectively terminate a pregnancy affected by Down syndrome, in chapter 6 I address the argument that there is nothing morally wrong with selective termination. Rather than the claim that there is a moral obligation to selectively terminate, this chapter examines whether there is a moral permission to selectively terminate. I do not argue that selective termination is wrong in all cases. Nonetheless there is something morally disquieting about the high selective termination rates after prenatal diagnoses of Down syndrome. I contend that, because of its pervasiveness in our culture, bias against people with cognitive disabilities must have some influence at the population level over decisions whether to continue pregnancies when given a diagnosis that the fetus has Down syndrome. People with Down syndrome suffer as a result of bias against them. We can raise a legitimate moral objection to actions undertaken out of bias, including selective termination. Prenatal testing and selective termination are not morally innocent in all cases since they are likely to be influenced by attitudes that are harmful and discriminatory. For these reasons, prenatal testing and selective termination can be morally questionable acts, though they are not necessarily morally questionable in all situations.

The chapter ends with a discussion of the strength of the moral duty I am proposing throughout *Choosing Down Syndrome.* I argue not that pregnant women should give birth to children prenatally diagnosed with Down syndrome in all cases. Instead I argue that more prospective parents should welcome children with this condition into the world and into their families. Furthermore I argue that the most defensible means for determining whether the choice to selectively terminate a pregnancy is morally justified is to leave this choice in the hands of the pregnant woman herself.

Chapter 7 compares the medicalized view of Down syndrome with how people with Down syndrome would like to be understood. The medicalized understanding of Down syndrome fails to acknowledge the extent to which social norms and social oppression contribute to being disabled. I defend a position on the status of Down syndrome that acknowledges the social causes of cognitive disability. I argue that we should be pragmatic about how we attribute "disability" to those diagnosed with Down syndrome. Though it is important in some situations to focus on the fact that a person with Down syndrome has a disability, in others, we should avoid regarding such a person through the restrictive prism of the person's disability. Ethical reflection should be employed when determining the appropriateness of attributing disability to someone. Foremost in our minds should be whether attributing disability to a person enhances that person's well-being or diminishes it. I take this normative pragmatic account of characterizations of disability and apply it to prenatal decisions about fetuses diagnosed as having Down syndrome. In the previous chapter, I introduce Adrienne Asch and David Wasserman's concept of the "sin of synecdoche" which is the failure to recognize the full humanity of the fetus with a Down syndrome diagnosis, effectively reducing the whole identity of the fetus to the single characteristic of having a disability. Chapter 7 considers the ethical justifiability of taking this perspective on the fetus.

The last chapter of the book outlines the political and economic ideology that influences, at various points, the decision to undergo prenatal testing for Down syndrome and the high rates of selective termination. The development of new prenatal tests to detect Down syndrome follows a capitalist imperative of innovation that involves creating new markets and shaping the desires of consumers. The medical professional has been coopted into the development of this market. Furthermore, many of the worries about bringing a child with Down syndrome into the world derive from preoccupations with making money and fears that such children are at a disadvantage in our capitalist market system. These worries are reflected in arguments (such as the "wrongful disability" argument) that make a case for selective termination by focusing on the diminished economic opportunities of people with disabilities as the primary factor that should guide reproductive decisions. At the same time, neoliberal economic policies and the dismantling of the welfare state have forced families into economic situations in which having a child with a disability seems like an overwhelming challenge.

From the perspective of an ideology in which what matters most is the perpetuation of the political and economic system of capitalism, prenatal testing and selective termination is a perfect process, alongside a perfect set of cultural expectations and beliefs about the worth of children, for eliminating human differences that do not fit well within a capitalist market system. From this perspective, prenatal testing and selective termination can be seen as a process whose intended outcome is the standardization of bodies and brains. The purpose of such standardization is the production of people who can contribute as workers in the economic system. The purpose of exposing these roots of the common decision to test and selectively terminate is to ask the reader whether he or she shares these political and economic values. Do you in fact share the values of this economic and political ideology? If you do not, then perhaps there is reason to make a different choice, the choice to bring a child with Down syndrome into the world, a child who can make different and worthy contributions to our social lives.

2 Ongoing Debates about the Ethics of Prenatal Genetic Testing for Down Syndrome

Ethical debates about prenatal testing and selective abortion have been with us for decades. The introduction of NIPT into prenatal care is a new development, however, with new implications for these debates. This chapter places it into the context of a number of issues and arguments that have shaped the debates in the recent past.

The chapter's first section addresses the question of whether noninvasive prenatal genetic testing will increase the number of selective abortions. Answering this question is important since, from the standpoint of a disability critique, the use of prenatal testing is only controversial because it leads to the elimination of people with disabilities through selective abortion. The second section introduces the "expressivist" objection to prenatal testing and selective abortion and outlines my position in relation to it. The third section discusses the debate about the impact of disabilities like Down syndrome on well-being. My argument in *Choosing Down Syndrome* crucially depends on empirical findings about well-being and family functioning, so it is important to outline how I understand the concept of well-being.

The fourth section considers prominent analogical arguments in the literature about the ethics of selective abortion: comparisons with selecting against a given gender ("sex selection"), preventing spina bifida with folic acid, selecting against homosexuality, and with the history of selective nontreatment of infants with disabilities. It concludes with a discussion of the current debate about the implementation of NIPT: whether these tests should be used as a form of public health screening.

The chapter has two goals. The first is to articulate my position in areas of long-standing and current controversy. Doing so will illustrate my overall argument in the rest of the book. The second is to provide essential background information for anyone who is new to the issue of prenatal testing and selective abortion. Renewed ethical interest could be generated by NIPT, but this new interest should be informed by important precedents.

Evidence for Increased Selective Abortion after Noninvasive Prenatal Testing

The impact of NIPT on selective abortion rates is currently unknown. There is some reason to believe that its greater use will result in an increase in the number of selective terminations. There is already some evidence from the United Kingdom that selective terminations after a diagnosis of Down syndrome have increased since the introduction of NIPT—693 selective terminations in 2014, compared to 512 in 2011 (Macfarlane 2015). Furthermore, the National Screening Committee and the Nuffield Council on Bioethics in the United Kingdom have both predicted that the overall number of selective abortions will increase with the planned introduction of NIPT as a secondary screening test to the United Kingdom's National Health Service (Nuffield Council on Bioethics 2017). This prediction is based on a pilot study involving eight hospitals and close to 1,000 women who were offered NIPT as a secondary screening test (Nuffield Council on Bioethics 2017). This issue requires more rigorous study. Some commentators are doubtful that widespread use of NIPT will result in an increase in selective abortion (see Michie 2016). In this section, I will outline my reasons for believing that selective terminations after a diagnosis of Down syndrome will increase.

The standard prenatal testing procedure prior to the introduction of NIPT, though still available to pregnant women, consists of a screening test such as maternal serum screening, followed by a diagnostic test, such as amniocentesis or chorionic villus sampling, if the screening test is positive. The screening test is often supplemented by, or combined with, other testing modalities such as nuchal translucency ultrasound or other ultrasound measurements. Rates of refusal of both screening and diagnostic tests differ depending on the population being studied, but many pregnant women refuse screening tests or refuse diagnostic tests after a positive screen. In a Dutch study, 62 percent of women refused a prenatal screening test, such as maternal serum screening (van den Berg et al. 2005). A similar American study documented that 18 percent refused (Markens et al. 1999). Refusals of amniocentesis after a positive screening result occurred 39–50 percent of the time in one study (Kobelka, Mattman, and Langlois 2009) and 30–40 percent in another (Mueller et al. 2005).

A number of studies have also investigated the reasons women give for refusing prenatal testing. These studies show that many women refuse prenatal testing because of problems or risks associated with the available tests. For example, Rayna Rapp's (1998) anthropological study of amniocentesis shows that the most common reason for deciding against the test was fear of miscarriage. In a study conducted with Dutch women, 32 percent of the women who refused maternal serum screening stated that adverse characteristics of invasive diagnostic tests such as the possibility of miscarriage after amniocentesis was one of the reasons for refusing. A study with American women documented that 16 percent of refusers of maternal serum screening did so because a positive test would lead to amniocentesis (Markens, Browner, and Press 1999). The same study showed that 44 percent refused because maternal serum screening is inaccurate—meaning that the test gives an odds ratio rather than a diagnosis that the fetus has a condition like Down syndrome (Markens, Browner, and Press 1999). Similarly, in the Dutch study, 42 percent of refusers were opposed to serum screening because of unfavorable characteristics of the screening test, such as its inaccuracy and the likelihood that test results could cause unnecessary anxiety (van den Berg et al. 2005).

Both the American study and the Dutch study suggest that refusal to undergo prenatal testing is not always motivated by opposition to abortion. In the American study by Susan Markens, C. S. Browner, and Nancy Press (1999), 64 percent of women who refused prenatal testing indicated that they refused because the initial screening test is linked to abortion, or because they would not terminate in any case. In the Dutch study conducted by Matthijs van den Berg and colleagues (2005), 15 percent of women refused because of opposition to abortion. As we can see from these figures, many women in both studies refused prenatal testing simply because the available tests were perceived to be risky or deficient, not because they were opposed either to prenatal testing or to abortion (van den Berg et al. 2005; Markens, Browner, and Press 1999).

Since noninvasive prenatal genetic testing is a simple maternal blood test that will not cause a miscarriage, and since it is more accurate than maternal serum screening, NIPT rectifies many of the problems motivating

Table 2.1
Reasons for and rates of refusal of prenatal testing

Refused screening test	Refused amniocentesis after screen positive test	Refused screening because could lead to amniocentesis	Refused screening because inaccurate; other deficiencies	Refused screening because opposed to abortion, will not abort
62% (van den Berg et al. 2005)	39–50% (Kobelka, Mattman, and Langlois 2009)	32% (van den Berg et al. 2005)	42% (van den Berg et al. 2005)	15% (van den Berg et al. 2005)
18% (Markens, Browner, and Press 1999)	30–40% (Mueller et al. 2005)	16% (Markens, Browner, and Press 1999)	44% (Markens, Browner, and Press 1999)	64% (Markens, Browner, and Press 1999)

refusal of prenatal testing with previously available tests. The availability of NIPT could consequently increase the use of prenatal testing. The women who were likely to refuse testing because of the inadequacies of previous screening or diagnostic tests, but were not opposed to abortion, are those whose use of NIPT could mean an increase in the overall numbers of pregnant women who undergo prenatal testing. Even when used as a primary or secondary screening test, NIPT could lead more pregnant women to enter the cascade of prenatal testing. Furthermore, the increased accuracy of NIPT could lead more women whose screening test results were positive to take on the risk posed by a diagnostic test such as amniocentesis. A pregnant woman who is fairly certain that she is carrying a fetus with Down syndrome might be more likely to undergo an amniocentesis she would previously have refused because of a less certain screening test result.

NIPT can be administered in the first trimester of pregnancy, which can shorten the timeline from screening test result to diagnostic test. Consequently, women may know earlier in pregnancy that they have a fetus with Down syndrome. From a technical standpoint, an earlier abortion is easier to perform than a later one. For pregnant women, an abortion earlier in gestation might be easier to undergo. Earlier gestational age is associated with higher termination rates after prenatal diagnosis of Down syndrome (Natoli et al. 2012). Again this analysis is speculative. However, for these reasons we can predict that selective abortion rates for prenatal diagnosis of Down syndrome will stay in the 60–90 percent range, or might actually increase, which means that greater use of prenatal testing in the form of NIPT could result in a greater overall number of selective abortions of fetuses with Down syndrome.

Nonetheless, even if this prediction does not come to pass, the argument that I have presented in this book would be unaffected. To be sure, the number of selective abortions and the rate of selective abortions might remain stable. But the case I make for choosing to bring children with Down syndrome into the world still applies even if the rates of selective abortion remain at their currently high levels. If the widespread use of NIPT should somehow correspond to a drop in selective abortion rates, this might indicate a greater acceptance of people with Down syndrome in our communities. Such a welcome turn of events would suggest that some of my claims about bias against people with this condition (found in chapter 6) do not hold true. However, most of the reasons I give for choosing to have a child with Down syndrome would still apply in such a scenario.

The Expressivist Objection

One of the most long-standing areas of debate about the ethics of prenatal testing and selective abortion centers around a position known as the "expressivist objection." This objection is the claim that prenatal testing and selective abortion are morally objectionable because they send negative messages about the lives of people with disabilities—that their lives are not worth living, or that such people are not worth including in our families (Asch and Wasserman 2005). Marsha Saxton writes that, "The message at the heart of widespread selective abortion on the basis of prenatal diagnosis is the greatest insult: some of us are 'too flawed' in our very DNA to exist; we are unworthy of being born" (Saxton 2010, 131).

Commentators such as Jamie Lindemann Nelson (2000) and Eva Feder Kittay (Kittay and Kittay 2000) argue that proponents of the expressivist objection have not convincingly established that nonlinguistic actions such as prenatal testing and selective abortion convey a definite linguistic message. To the extent that prospective parents may choose to test and selectively abort for reasons innocent of bias against disabilities, there is no necessary connection between these actions and negative messages purportedly arising from them.

The disability critique I advance is not based on an objection to the messages sent by selective termination or prenatal testing. I do not understand my critique to be an expressivist objection. Nonetheless, at various moments in my argument, my analysis does focus on the motives underlying choices to test and terminate. A number of authors seem to assume that such an analysis of motives is the same kind of critique as the expressivist objection—notably Jonathan Glover (2006) and Stephen Wilkinson (2010). Both Glover and Wilkinson agree that the expressivist objection is a legitimate critique of selective termination and prenatal testing when these actions are motivated by biased attitudes against people with disabilities. Glover (2006, 35) calls these "ugly attitudes." Wilkinson (2010, 185) calls them "unduly negative attitudes." Glover and Wilkinson both believe that people ought neither to hold these attitudes nor to act on the basis of them. They question, however, whether these biased attitudes are actually a motivating factor in people's reproductive decisions. They take issue with the expressivist objection because they do not believe that people tend to be motivated by attitudes of bias in their reproductive decision making.

Wilkinson (2010, 174) says that, for the most part, prospective parents seem to act "on the principle that health is better than disease or disability (presumably on child welfare grounds ...)." Prospective parents also choose to avoid having children with disabilities because of concerns that raising such children will make it more difficult to realize their own career goals and personal projects (Wilkinson 2010, 180). For his part, Glover claims that the expressivist objection can be dispelled if we reassure people with disabilities that our reproductive decisions are not motivated by ugly attitudes. We can do so by taking concrete actions such as ensuring nondiscrimination, accessibility, and inclusion in social supports, housing, and employment (Glover 2006, 33–35).

I agree with Wilkinson and Glover that a disability critique of prenatal testing and selective termination can involve an analysis of motives and attitudes that underlie these actions. But I disagree that we can be easily reassured that decisions about prenatal testing and selective abortion are not usually motivated by antidisability bias. These authors assume that our motives are ultimately transparent to ourselves and that deciding whether to bring children with disabilities into our families tends to be untainted by bias (see Wilkinson 2010, 181). I argue (particularly in chapter 6) that it is highly likely that our reproductive decisions are influenced by unacknowledged or disavowed cultural prejudices about disabilities and people with disabilities. On a population level, these "ugly attitudes" must play a role in the high rate of selective termination for Down syndrome, even though people rationalize their choices in terms of more palatable concepts such as concerns about the baby's "health" or concerns about one's own career.

Well-Being

Given that my case for bringing children with Down syndrome into the world involves the claims that the lives of such children tend to go well, and that their families function well, the concept of well-being is crucial for my argument.[1] The impact of disabilities like Down syndrome on well-being is a contested topic in philosophical ethics (Bickenbach, Felder, and Schmidt 2014). My understanding of the concept of well-being is simple: I take subjective assessments of well-being by people with Down syndrome (and other disabilities) to be authoritative almost all of the time. As I show in chapters 3 and 4 and elsewhere, there is substantial empirical evidence documenting that people with Down syndrome and other disabilities tend to positively assess their own well-being (Skotko, Levine, and Goldstein 2011c; Albrecht and Devlieger 1999; National Organization on Disability 1994). There is also substantial empirical evidence showing that families that include a person with Down syndrome tend to function well (detailed analysis and citations in chapter 3). This data involves both subjective and objective measures. There is also a large body of anecdotal evidence supporting both claims, some of which is presented in chapter 3.

Other writers on the topic of disability and well-being do not take subjective assessments to be authoritative. As I show in chapter 5, the authors of the influential book *From Chance to Choice* (Buchanan, Brock, Daniels, and Wikler 2000) believe that there is an objective component of well-being that is not captured by subjective assessments. For these authors, the main objective determinant of well-being is "opportunity." Dan Brock explains in another work that "opportunity for choice among a reasonable array of life plans is an important and independent component of quality of life" (Brock 1993, 124). Subjective assessment of one's well-being may miss the degree to which opportunity is available in one's own life. Brock and his coauthors run into problems defending this claim, however. As I detail in chapter 5, in the place of argument, they offer only intuitions about opportunity as an essential component of well-being—when such intuitions have been contested by people with disabilities and their advocates.

Jonathan Glover acknowledges the evidence that people with disabilities often report that they experience high levels of well-being. However, he also searches for an objective understanding of well-being that can allow us to question these self-assessments. Glover seizes on the Aristotelian idea of human flourishing, which lends itself to objective interpretation (Aristotle 1984). In Aristotelian terms, we flourish as human beings when we function well in accordance with our essential human capacities—which include rationality and sociability (Aristotle 1984). Even though we might think we are doing well, we might not be flourishing if we are not very good at exercising our rationality, for example, or if we have limited social lives. If a disability is a functional limitation that affects capacities essential for our humanity, such a disability could inhibit our flourishing.

The problem for Glover's account, however, is that in our nonessentialist era, he cannot easily utilize an Aristotelian account of human nature or essence to determine which capacities are necessary for human flourishing. He admits that "there is no one recipe for human flourishing" (Glover 2006, 15). Martha Nussbaum has a detailed Aristotelian account of human flourishing (her "capabilities account") that does not cast people with cognitive disabilities as limited in their ability to flourish (Nussbaum 2006). Because of differences in vision about what human flourishing requires, Glover defers to subjective appraisals of this concept, including the opinions of people with disabilities (17). But he thinks these opinions require "interpretation"—the self-assessments made by people with disabilities about whether they are flourishing should not be taken as authoritative. Glover's account on this topic ends on a circular and indecisive note when he points out that our interpretations of opinions about human flourishing should, in part, be based on "the history of reflection on the components of human flourishing" (23). The most notable figure in this history of reflection is, of course, Aristotle—whose views require supplementation by self-assessments of people with disabilities, and so on, around the circle.

Wilkinson has a similar account. His point of departure is the assertion that empirical survey research to determine the quality of life of people with different disabilities is unnecessary or beside the point because "there is an *a priori* connection between disability and welfare" (Wilkinson 2010, 63). He follows Glover in locating this connection in the concept of human flourishing. He says, "characteristics only get to *count* as disabilities, as opposed to mere differences, if they impair the capacity to flourish" (Wilkinson 2010, 63). However, Wilkinson does not get any further than Glover, finding that whether a disability inhibits the capacity to flourish is highly context dependent. The link between disability and lower quality of life is, he says, "murky" (Wilkinson 2010, 68). He admits that it is "inadvisable to generalize about the relationship between disability ... and quality of life" (68).

Nonetheless, Wilkinson is undeterred from making such generalizations himself. He grants "for the purposes of discussion" and without argument, the claim that "selecting for disability means selecting a lower quality of life" (Wilkinson 2010, 68)—in other words, a decision to give birth to a child with a disability like Down syndrome will necessarily result in giving birth to a child with a lower quality of life than a nondisabled child. Though these positions taken by Brock, Glover, and Wilkinson are poorly justified, they are tenaciously held, despite their flaws and deficiencies.

I have outlined three philosophical positions that attempt to overrule, on "objective" grounds, the subjective assessments of people with disabilities about their well-being. These three positions fail to establish an objective standard of well-being that will stand in opposition to the empirical findings on disability and self-assessed well-being. Among the philosophers who think that disabilities negatively affect well-being, Brock and colleagues, Glover, and Wilkinson are actually more reflective than others. Jeff McMahan, for one, seems to assume that there is a direct correlation between intelligence and well-being, such that people with cognitive disabilities will necessarily have lower well-being than people without such disabilities (see, for instance, McMahan 2002, 244). Peter Singer appears to make similar assumptions, claiming without argument that "a normal child" will have "a better life" than a child with Down syndrome (Singer 1993, 187–188). Rather than directly challenging the empirical findings about disability and well-being, neither McMahan nor Singer actually engages at all with these findings in these works. The assumption that higher intelligence necessarily leads to higher degrees of well-being is simplistic, unsupported by evidence, and far too contentious to be accepted without argument.

Julian Savulescu and Guy Kahane (2009, 286), for their part, take the link between disability and diminished well-being to be so close that they actually *define* disability as a state that lowers well-being. (I examine their arguments in greater detail in chapter 5). In advancing their "Principle of

Procreative Beneficence," Savulescu and Kahane (2009) demonstrate some awareness of the empirical support for the idea that people with disabilities exhibit levels of well-being comparable to those of nondisabled people. But, as I demonstrate in chapter 5, their account is rife with undefended and untenable assumptions about the nature of well-being.

Because of this list of failures to discredit the empirical findings on the well-being of people with cognitive disabilities, I see no reason to question the findings myself. First of all, the burden of proof lies on those who would overrule the subjective reports. The philosophers who think that cognitive disability necessarily leads to diminished well-being have not met this burden. Instead of convincing argument they offer evasions, biased intuitions, circular arguments, and untested assumptions. Second, it is unseemly to cast doubt on the empirical findings about disability and well-being without good reason (Goering 2008). The empirical evidence is extensive and convincing that people with Down syndrome typically experience a degree of well-being comparable to people without disabilities. The alternatives to these empirical findings are weak. To favor the assumptions and poor arguments of philosophers over the considered testimony of people with Down syndrome themselves—reflecting about their *own* lives—would be the height of epistemic arrogance.

Third, if we consider a plausible catalog of the things that make someone's life go well, we find that these things tend to be within reach of people with Down syndrome and other disabilities. Nussbaum's capabilities account provides a list of ten human capabilities that she argues are necessary for a flourishing life. She lists life; bodily health; bodily integrity; the use of senses, imagination, and thought; the emotions; practical reason; affiliation; interaction with other species; play; and control over one's environment as the ten categories of necessary capabilities for human well-being (Nussbaum 2011). Each capability must be secure for a person above a minimum threshold for that person's life to go well. Applying her account to some examples of people living with Down syndrome and autism, Nussbaum argues that "there is a realistic prospect that [these people] will attain the capabilities that we have evaluated as humanly central" (193). Nussbaum's capabilities account lends philosophical support to the empirical findings about disability and well-being.[2]

Earlier, I stated that I accept subjective self-assessments of well-being by people with disabilities to be authoritative almost all of the time. The exceptions would not apply to any of the empirical sources that I cite and discuss, but would have to be acknowledged in a more general theory of well-being, which I am unable to offer here.[3] "Adaptive preference" is a phenomenon in which those living in oppressed or impoverished circumstances give high subjective quality of life assessments (Khader 2011). This issue comes up in chapter 4. In such circumstances, there might be reason to doubt self-assessed well-being. My position is that such doubts are warranted when a person's external social circumstances are so bad—because of abuse, violence, starvation, deprivation—that their self-assessment of well-being seems radically dissonant. There is no reason to believe that these conditions obtain for the research participants in the studies about well-being and Down syndrome that I cite or for the authors whose testimonials I cite.

Common Analogies in the Debate

In the ethical debate about prenatal testing and selective abortion for Down syndrome, philosophers and bioethicists on both sides often draw on analogies and comparisons to defend their positions. These comparisons are so obvious and prominent in the literature that they merit mention—and they also help to illustrate arguments supporting my position.

Sex Selection

Several authors compare the termination of fetuses with disabilities to abortion for the purposes of sex selection (Asch 2000; Saxton 2000). Here I will refer to selective abortion after prenatal diagnosis of a disability as "selecting against disabilities," and I will refer to abortion for the purpose of selecting out female fetuses as "sex selection."

Adrienne Asch, for instance, points out that many people oppose the use of prenatal testing and abortion for sex selection (Asch 2000). For example, the legal scholar Jeffrey Botkin's (2000) model of professional standards for prenatal diagnostic services and the President's Commission for the Study of Ethical Problems in Medicine and Biomedical Research (1983) both recommend that prenatal testing and selective abortion not be used for sex selection. Asch (2000) argues that opponents of sex selection should oppose selecting against disabilities for the same reasons. Sex selection is motivated by a view of the fetus in which one characteristic (sex) is considered determinative of who that future person will be. According to Asch, this reductive view is a form of discrimination. Opposition to sex selection is based on opposition to discrimination. Those who oppose sex selection should be opposed to selecting against disabilities because each is equally discriminatory. Marsha Saxton (2000) also argues that sex selection and selecting against disabilities are similar. For Saxton, both are the results of oppression—the oppression of women and the oppression of people with disabilities.

One way of countering these arguments is to argue for the acceptability of sex selection. Some writers who support selecting against disabilities use this strategy (see Wilkinson 2010, 31; Steinbock 2000, 123). Another possible response is to argue that the two practices are different in relevant ways. Bonnie Steinbock adopts this strategy by suggesting that, unlike sex, disability is a health issue and selecting against it is a form of prevention. Steinbock argues that preventing disability in one's future child through selective abortion is like taking folic acid to prevent spina bifida in a future child (Steinbock 2000, 118). Steinbock claims that sex is not a health category like disability. It makes sense to take preventive measures to preserve the child's health. Avoiding disability is a way of preserving health. Those adopting this strategy argue that sex selection and selecting against disability are different.

I will discuss the comparison of selective abortion for Down syndrome and folic acid supplementation to prevent spina bifida below, as well as some of the ethical aspects of viewing selective abortion as a form of prevention. But Steinbock's most contestable claim is that disability means ill health. The link between the two concepts is not very strong for disabilities like Down syndrome or deafness because someone who has Down syndrome or who is deaf can be perfectly healthy. If we contest the idea that avoiding disability is a way of preserving health, then the distinction Steinbock puts forward between selecting against disability and sex selection seems not so significant. Asch's argument that both forms of selection are motivated by discriminatory attitudes appears more compelling.

The comparison with sex selection plays no role in my argument in the rest of the book. Nonetheless, the comparison is compelling because of parallels with some of the feminist arguments against sex selection. Many people with feminist commitments object to sex selection as it is practiced in places like South Asia. Feminists often object to sex selection because of the discriminatory motives that underlie the practice, and because of the widespread social discrimination that is its context (Bubeck 2004). Selective abortion against Down syndrome (so I will argue) is also motivated by discriminatory attitudes. People with Down syndrome pursue their lives in an atmosphere of social discrimination, (although this atmosphere is likely not as poisonous as that experienced by women in many places in South Asia). I find Asch's comparison compelling, though I do not treat this argument as conclusive. If you oppose sex selection because it is a manifestation of bias, you should also oppose selecting against disability.

Preventing Spina Bifida with Folic Acid

As mentioned above, some authors argue that selective abortion is similar to taking folic acid supplements during pregnancy to prevent spina bifida (Steinbock 2000; Malek 2010). The disability critique of selective abortion is not based on the premise that fetuses are harmed by terminating a pregnancy. If there is nothing morally troubling about abortion, then the act of terminating a pregnancy can be portrayed as akin to other actions that prevent one's future child from having a disability. According to this argument, if it is morally innocent to prevent a disability by taking folic acid, then it is morally innocent to selectively terminate for Down syndrome.

As I show in chapter 5, similar reasoning leads to arguments that selective termination to prevent disability is morally *required,* as are adopting a healthy diet during pregnancy and taking nutritional supplements to avoid disabilities. As I see it, the comparison between selective termination and folic acid supplementation overlooks a morally relevant difference between these two actions. The difference can be seen when one compares the treatment of a disability with efforts to prevent the birth of a future child with a disability. Correcting a disability through treatment is consistent with accepting the child with the disability, whereas preventing the birth of a child with a disability is not consistent with this attitude of acceptance. Folic acid supplementation is more like treating a child than preventing the birth of a child because supplementation is consistent with an attitude of acceptance if the preventive efforts fail.

In the case of selective abortion, a previously wanted pregnancy changes to an unwanted pregnancy on the basis of a prenatal diagnosis. There is no such change when pregnant women supplement their diet with folic acid. As I argue in chapter 7, selective abortion is a refusal to enter into a parenting relationship with a prospective child who will have a disability. There are many reasons for this choice, but such a refusal is an attitude of rejection toward that prospective child. Legitimate moral objections may be raised against such attitudes because of the common phenomenon of discrimination against people with disabilities like Down syndrome. I argue that discriminatory attitudes must play a role, at the population level, in the high rates of selective termination. Folic acid supplementation does not have this necessary feature of refusal and rejection. A pregnant woman may supplement her diet with folic acid in order to prevent disability and yet, should that effort fail, be fully prepared to welcome a child with that disability into her family.

Selecting against Homosexuality

Michael Sandel relates a story about James Watson, the scientist who helped discover the structure of DNA. Sandel (2007, 71) writes, "Watson had stirred controversy by saying that, if a gene for homosexuality were discovered, a pregnant woman who did not want a homosexual child should be free to abort a fetus that carried it.... His remark provoked an uproar." If one considers the grounds for objecting to selective abortion for homosexuality, there are certain similarities with selectively aborting fetuses that have Down syndrome. Watson might point out that people with pro-choice convictions cannot object on the basis that gay people are harmed by such choices since selective abortion would not (obviously or directly) affect gay people already born. Nonetheless, bias against gay people persists, and we can infer that decisions to abort fetuses "diagnosed" as homosexual would be motivated, in many cases, by homophobia. One could thus object to prenatal diagnosis and selective abortion of "homosexual" fetuses as part of an objection to the common bias against gay people in our culture. The nature of this objection is similar to that of the objection against selective abortion for Down syndrome that I am defending.

In his book *Ethics, Sexual Orientation, and Choices about Children.* Timothy F. Murphy (2012) considers the arguments in this (currently) hypothetical debate about prenatal efforts to prevent the birth of children who are gay, and the research that could lead to the development of tests that enable these choices. Murphy defends a parent's right to choose to prevent the birth of children with an undesired sexual orientation. One argument in support of this position is that letting parents make these choices would avoid bringing children into families hostile to their sexual orientation, families who would ultimately harm the children (Murphy 2012). It would be perverse to require parents to bring children into the world whom they would ultimately harm. This argument is fairly, though not totally, convincing. I examine similar motivations for selective abortion of fetuses with Down syndrome in chapter 7.

One deficiency of this argument is that it does not take into account people's capacity to change. The parenting memoirs I examine in chapter 3 show many examples of parents who previously did not want a child with Down syndrome but who now cherish and welcome such a child. Often it is the birth of the child that is a catalyst for a change in attitudes and values. These examples of moral progress are positive developments for the acceptance of people with Down syndrome in our society. Similar progress might occur when previously homophobic parents find out that their children are gay. Parents' acceptance of their gay children can contribute to increasing acceptance of being gay in our society. Of course, not all homophobic parents will be able to let go of their bias when their gay children "come out"—so this argument is not strong enough to support a moral requirement to forgo prenatal testing and selective abortion for Down syndrome or homosexuality. But it is strong enough to support the claim that more people should welcome such children into the world.

Selective Nontreatment of Infants with Disabilities

Back in the 1970s and 1980s, it was common for parents and doctors to refuse lifesaving treatments for newborns with Down syndrome, which meant they would then die (Weir 1984; Strong 1984). Infants with Down syndrome are sometimes born with gastrointestinal malformations or heart conditions that must be surgically corrected in order for the baby to continue living. The infamous case of Baby Doe in 1982, discussed in the preface, involved an infant with Down syndrome who had an esophageal atresia who was denied surgery by his parents, and died six days after his birth from deliberate starvation and dehydration (Munson 2008, 636). In cases involving nondisabled infants, surgical correction of such a malformation would have been routine. The philosopher Peter Singer and others have defended the practice of killing newborn infants with disabilities (Singer 1993; Kuhse and Singer 1985).

I note this issue not to suggest that selective termination is similar to infanticide. Instead, I would like to compare the influence of research findings about family functioning on the selective nontreatment debate versus the influence of these findings on the more recent discussion about the ethics of selective termination.

During the 1970s and 1980s, research into families with children who have disabilities such as Down syndrome often found that such families were typically dysfunctional, and that the cause of dysfunction was the child's disability. Since then, this area of research has undergone significant changes in its assumptions and methodologies, as have the social circumstances of the families being studied. Current research findings show that families with a child who has Down syndrome typically function well. Curiously, however, the current evidence of good family functioning is not given the same weight in the selective termination debate as previous findings of poor family functioning were in the selective nontreatment debate. Researchers Philip M. Ferguson, Alan Gartner, and Dorothy K. Lipsky (2000, 86) point out that: "Twenty years ago, when the research on families predominantly seemed to support the medical predictions that such disabilities were unmitigated and chronic tragedies for the parents involved, the bioethical discussions cited this evidence at length. Now that the weight of family research has challenged this assumption, the evidence is seldom discussed. Instead, the claim is made that the issue all along has been parental choice." In other words, when previous empirical research suggested that families were negatively affected by a child with a disability, this research was used to support arguments that favored denying life-sustaining care for infants with disabilities such as Down syndrome. When the empirical research improved and began to show that such families actually function well, these conclusions were ignored. In this context, Ferguson and colleagues encourage us to think of the reasons given in these debates not just as arguments, but also as the products of cultural beliefs about disabilities. These beliefs are like prisms through which we see the lives of people with disabilities and their families.

In our culture, there is a pervasive tendency to believe that all disabilities are undesirable (Kent 2000). While under the influence of such a belief, if confronted by evidence that it is possible, even likely, to have good quality of life while living with Down syndrome, or that the families of such children are thriving, there is a tendency to either disbelieve the evidence or regard it as irrelevant. As Mary Ann Baily (2000, 70) admits in her essay justifying her reproductive choices: "Most people are not analytical thinkers; they tend to know what they want to do and then look for justifications." It is possible that many academic participants in this debate similarly let their conclusions precede their justifications. This possibility is perhaps not surprising given that our political beliefs, allegiances, and cultural identities influence the arguments we devise to support our beliefs. These influences are one reason for slowing the debate down. We must examine our premises and inferences, and ensure that our arguments are not the products of bias or discriminatory attitudes. If we draw attention to the way that these attitudes can motivate reproductive decisions, and can influence our thinking, then it is perhaps possible to overcome their influence. Such self-awareness could contribute to the social fund of knowledge about disabilities, and convince prospective parents that it is possible and desirable to bring a child with Down syndrome into the world.

Noninvasive Prenatal Testing as Public Health Screening

With the ongoing expansion of fetal genetic information available prenatally through NIPT, some authors have questioned whether publicly funded mass prenatal screening programs that currently exist in many countries should be maintained, or adopted by other countries (Munthe 2015; DeJong and DeWert 2015). Given the veritable flood of conditions that are testable prenatally, public health decision makers have the difficult task of choosing which conditions warrant inclusion in such screening programs. By making such choices, the expressivist objection would rear its head once again. The obvious implication to be drawn from including a condition within a screening program is that the birth of people with the condition in question should be avoided for the benefit of public health. The "official" nature of such a public health designation would eliminate most of the ambiguity of messaging that is at the root of some critiques of the expressivist objection (e.g., Nelson 2000). Even before NIPT, the inclusion of testing for Down syndrome and other conditions in public health screening programs was already a target of the expressivist objection (Juth and Munthe 2012).

Because of the inherent difficulty of running a mass screening program for the large number of testable conditions that are or will become available through NIPT, it is likely that screening programs will not be expanded. If so, prenatal screening and testing will become more and more individualized. Respect for the autonomous choices of prospective parents will likely become the guiding norm of offering NIPT, rather than a population-based norm of advancing public health.

The companies offering NIPT tend to market their services directly and aggressively to consumers (Munthe 2015). In order to use their proprietary tests, maternal blood samples must be sent off to centralized laboratories operated by the testing companies. In the United States, these NIPT services have been brought to market without regulation by the Food and Drug Administration (FDA), which would normally regulate new drugs, devices, and diagnostic tests (Daley 2014; Agarwal et al. 2013). The lack of regulation by the FDA is due to a regulatory loophole exploited by the testing companies (Daley 2014). FDA regulation has not historically extended to laboratory-developed tests, and the testing companies have presented their NIPT products as tests of this sort (Agarwal et al. 2013). Tests manufactured and sold for use in multiple laboratories are subject to FDA oversight. But tests that are designed, manufactured, and used within a single laboratory are not. In the case of NIPT, thousands of physicians send samples for testing at individual laboratories administered by the testing companies. This centralized model for the use of NIPT enables testing companies to avoid FDA regulation. Perhaps because of this lack of regulation, the new tests have been found to be less accurate than has been suggested by the testing companies (as was mentioned earlier; Meredith et al. 2016; Daley 2014).

Because of the complexity of the information generated by these new tests and problems with their accuracy, pregnant women and their partners need intensive genetic counseling to assist with reproductive decisions when given a "positive" test result. But genetic counseling is often unavailable, and the physicians conveying information to pregnant women are often neither adequately trained nor adequately informed to provide such counseling (Meredith et al. 2016). The need for intensive genetic counseling is another reason NIPT is moving away from a public health mass screening model toward a more individualized and commercialized one. As Christian Munthe (2015) observes, in many countries intensive genetic counseling was recommended and needed for participants in previous prenatal testing programs, but the supports provided were rarely adequate. An expansion of tests in a mass screening program will only increase the demand for genetic counseling. If such a demand cannot be met, it is ethically irresponsible to create such a screening program.

In the case of Down syndrome, my position is that this condition is not so serious that a publicly funded mass screening program would be warranted. The medical historian Ilana Löwy (2014) argues that de facto, if not official, mass prenatal screening programs targeting Down syndrome already exist in many countries as a result of advocacy on the part of physicians who believe that eliminating this condition is a public health imperative. But because people with Down syndrome tend to enjoy their lives (as I will argue in greater detail in later chapters), public health screening for Down syndrome using NIPT would be a misuse of resources. Recognizing the reality that NIPT is available to women for prenatal care, it should be offered to them on an individualized basis, respecting their autonomous wishes. In order to fulfill their responsibilities toward prospective parents, health care organizations should have sufficient genetic counselors or sufficiently trained physicians and midwives available to help these woman and their partners interpret results of NIPT and support reproductive decisions.

Though I favor an autonomy-based model over a public health screening model for offering NIPT, each of these models is susceptible to the disability critique. Stephen Wilkinson (2015) makes this point convincingly. As long as antidisability bias is a cultural factor influencing high selective termination rates for Down syndrome, the disability critique can be leveled as much at an autonomy-based model as it can at a public health screening model. Though if forced to choose between the two models, the autonomy-based model is more ethically justified— because under this model, autonomous refusals of testing or refusals of selective termination on a large scale could result in a situation in which the disability critique is no longer needed. If selective abortion rates dramatically decrease, then the disability critique of prenatal testing for Down syndrome will have reached its goal.

3 Autobiographical Accounts of Parenting a Child with Down Syndrome

In her memoir *Bloom*, Kelle Hampton writes that when her daughter Nella was born, she immediately suspected that Nella had Down syndrome. Just as with any child, the birth was a momentous event. But a second momentous event took place soon afterward. Hampton writes, "Suddenly, my pediatrician walked in, and my heart sank. This was it. 'Why is she here?' I asked.... And it happened" (Hampton 2012, 8). The "it" that happened was the pediatrician's confirmation that Nella indeed had Down syndrome. A diagnosis of Down syndrome, either before or after the birth of your child pulls you across a threshold.

Other parents of children with Down syndrome tell similar stories about the change brought about at the moment of diagnosis. Renee Parker (2007, 269) relates that, when her daughter was diagnosed soon after birth, "those words seemed to cast a spell on everyone in the room." The spell usually touches off an intense wave of sadness. Parents often describe this sadness as a form of grief, akin to the loss of a child. But their sadness has an odd, indeterminate aspect, as though receiving this news truly places them under a spell. Hampton describes her sadness in great detail, as well as the circumstances and events surrounding Nella's birth, but admits that "I couldn't even tell you specifically what I was sad about" (Hampton 2012, 67). After the birth of her son Thomas, Kathryn Lynard Soper (2009b, 110) writes that she "had ample evidence that something *was* wrong" but felt very conflicted and unsure about what this something might be. Looking back on her daughter's prenatal diagnosis, Nancy Iannone (2007, 131) asks, "Why was I so upset?" and suggests that "there was not much considering going on. It was just sadness." Very often the story of having one's child diagnosed with Down syndrome begins with an inchoate sense of loss.

The cognitive deficits often associated with Down syndrome only become apparent later in childhood. So at birth or in the prenatal period, a baby or fetus with a Down syndrome whose diagnosis bestows the label of "disabled" does not typically exhibit disability.[1] A parent's sense of loss is brought about by a feat of imagining—of projecting the social meanings of Down syndrome into the baby's future. This lack of tangible deficit could help explain the indeterminate nature of a parent's sadness.

As I have argued, the use of new noninvasive prenatal testing techniques could lead to an increase in the selective termination of pregnancies affected by Down syndrome. Memoirs like the ones written by Hampton, Parker, Soper, and others provide some insight into why this increase in selective termination should not happen—and why more prospective parents should choose to have a child with Down syndrome. These narratives provide richly detailed accounts of what it is like to welcome a child with Down syndrome into your family. They testify that parenting a child with Down syndrome is an overwhelmingly positive and transformational experience.

Kelle Hampton's *Bloom* (2012) is a recent addition to the genre of book-length autobiographical accounts of parenting a child with Down syndrome through the first days and years of life. Included in this genre are Rachel Adams's *Raising Henry* (2013), Martha Beck's *Expecting Adam* (1999), Michael Bérubé's *Life as We Know It* (1996), George Estreich's *The Shape of the Eye* (2011), Jennifer Graf Groneberg's *Road Map to Holland* (2008), and Kathryn Lynard Soper's memoir *The Year My Son and I Were Born* (2009b). Soper is also the editor of two anthologies of short narratives: *Gifts* (Soper 2007), which consists of sixty-three autobiographical essays written by parents of children with Down syndrome, and *Gifts 2* (Soper 2009a), in which thirty-three of seventy-six short essays are written by parents, and the remaining forty-three by friends and other family members of people with Down syndrome. These book-length and short autobiographical parent narratives form the source material for this chapter.

When parents are told that their child has Down syndrome, they are forced to think about what this condition means for them, to confront their biases and presuppositions, and to evaluate society's messages about cognitive disability. The process of writing a memoir makes explicit parents' effort to understand the significance of the birth of a child with Down syndrome in their lives. Parenting memoirs show that the uncertainty and conflicted feelings brought about by a diagnosis of Down syndrome are a symptom of the transformative nature of this experience. Receiving news that your child has Down syndrome often places one on the cusp of a personal transformation, a reorientation of values. The parent narratives show many examples of people rooted in a system of conventional middle-class values that esteem achievement, competitive advantage, individuality, and independence who come to embrace a new understanding of what is important in life—values such as compassion, care for one another, unconditional love—as a result of parenting a child with Down syndrome.

Initial Reactions to the Diagnosis

The stories written by parents of children with Down syndrome detail a range of emotions that are touched off by finding out that their child has this diagnosis. Fear and worry about the child's future are mentioned frequently (e.g., Roberts 2007, 154; Castellano-Farrell 2007, 237; Beck 1999, 186). Two other responses stand out. For one, several authors admit that they fantasized about fleeing or escaping soon after hearing the diagnosis. Carey Branam (2007, 210) writes that "I constantly imagined how I could disappear and start over." In *Road Map to Holland,* Jennifer Graf Groneberg writes in the present tense, as though the events are unfolding before our eyes. She says that, after giving birth to twins, one with Down syndrome, "I have an overpowering desire to flee. To get up and run.... I could start over, be someone else, anyone but who I am" (Groneberg 2008, 5). Similarly, Kelle Hampton (2012, 17) painfully confesses that "I cried out that I wanted to leave her and run away. I wanted to take Lainey [her first child] and my perfect world and this perfect love I had built with my two-year-old and our cupcake baking days and our art projects and our beautiful bond and I wanted to run like hell."

A second common response is a profound sense of guilt. The authors feel guilty for a number of reasons: they feel responsible for their child's Down syndrome, they do not feel up to the task of raising a child with special needs, or they feel ashamed about their initial feelings of rejection toward the child. Jennifer Groneberg's guilt was wide ranging. She writes, "I have so much guilt. I was too old to have a baby. Or it's deeper within me, a rotten core of bad genes that Avery has to pay for.... I was so scared. He could be held by anyone; and so I let him. He was everybody's baby but my own" (Groneberg 2008, 227). Soper (2009b, 60) writes that "I couldn't shake my guilt over feeling grief to begin with." Among the most extreme reactions, Colleen J. Bailey (2009, 214–215) became suicidal because of her feelings of guilt. Bailey writes that "I sat alone and cold on a remote country road staring at all the prescription bottles I had grabbed before leaving the house.... I felt like a failure. I thought someone else would be a better mom." Bailey's grieving process, along with Soper's, was made less tractable than that of others because their grief was compounded by depression.

During the grieving process in many of these parent stories, the object of grief and worry is not so much the newborn child as it is the author himself or herself. Kathryn Soper (2009b, 21) realizes that "as much as I wanted Thomas to have a good life, right then I cared more about whether *I* could have a good life." Speaking of her grief over the prenatal diagnosis of her son Riley, Andi Matthews (2007, 287) admits that "I was primarily thinking about myself. I couldn't imagine how it would feel to see my own son teased." When given her son Adam's prenatal diagnosis, Martha Beck (1999, 184) wonders, "What does it mean to my life—to our lives?" Similarly, upon the birth of his son Jamie, Michael Bérubé (1996, 6) wonders, "Would we ever have normal lives again?"

In a way, the child with Down syndrome often fades into the background in these narratives, assuming the role of a minor character, and almost seems to disappear. The child's arrival causes emotional upheaval, but because of the passive nature of newborn existence and the indeterminate state of the child's much feared disability, most of the action revolves around the parents, siblings, other family members and friends, and strangers. We often realize these absences only when the child makes an appearance late in the narrative. Soper once again provides a good example. For much of the first months of his life, Soper's son Thomas was hidden by medical apparatus, oxygen tanks, and nasogastric (NG) tubing. Then, some 100 pages into her story, she tells us that "the NG tube was finally gone. His face looked so bare and clean.... His eyes, gray and brown all at once, peered up at me. *Here I am,* he seemed to say" (Soper 2009b, 112). Jennifer Enderlin Blougouras's metaphorical account compares having a child with Down syndrome with being thrown into the deep end of a pool. Blouguoras (2007, 63) describes how she flails around in the pool, wondering how this happened to her, and whether she could survive the experience. Her little boy Nicholas only appears at the end of the story, *after* she realizes that she can float (63).

These disappearances of the child are not central to the plots of all Down syndrome parent narratives. Martha Beck's Adam and Michael Bérubé's Jamie, for example, are characters drawn in rich detail who do not "disappear." But when these disappearances do occur, they highlight a central point about the experience of having a child with Down syndrome. Much of the time, Down syndrome is just a collection of ideas through which the newborn child is understood, rather than a concrete set of characteristics used by the author to flesh out the child's character. As a set of ideas, the meaning of Down syndrome is provided from the outside, by the wider culture. As the child recedes from the narrative into a passive role, the parent's hopes and fears take center stage, along with the values and images of the surrounding culture.

The Influence of Middle-Class Values

Down syndrome parent narratives often take place, at least in part, in hospitals. But, as the stories move into the days and months after the birth of the child, their settings are unmistakably within the geography of middle-class America. Kelle Hampton (2012, 252) uses the quintessential symbol of American middle-class life to describe the emotional impact of finding out that her daughter had Down syndrome—the white picket fence. She says, "once, a long time ago, I had a white picket fence, but it fell down" (Hampton 2012, 252). Many events in these stories take place while shopping. We encounter their characters in the aisles of Walmart (Guillaume 2007, 150; Spring 2007, 135), Kmart (Beck 1999, 73–75), and Target (Hampton 2012, 25, 151). They go for coffee at Starbucks (Estreich 2011, 52; Lester 2007, 254), Dunkin' Donuts (Hampton 2012, 62), and 7-Eleven (Hampton 2012, 226). One pregnant woman contemplates what to do about her fetus's prenatal diagnosis while at a conference in Vegas (Lester 2007, 255). They shop at Baby Gap (Hampton 2012, 30), rent movies at Blockbuster (Soper 2009b, 139), and go out for dinner at Outback Steakhouse (Huffman 2007, 41). Their emotional crises take place against a background of bourgeois normalcy.

The norms and images of middle-class life are often used in these stories as a way of expressing their sense of grief. In a moving passage, Kathryn Soper (2009b, 64) describes seeing a man with Down syndrome on a commuter train after she comes home from a long stint in the neonatal intensive care unit (NICU) with her son Thomas. She writes, "The man in the red jacket—that was Thomas. He would ride public transportation someday, having awkward conversations with unwilling seatmates as he headed to some menial job—if he was lucky." Within the context of car-centric suburban individualism, taking public transit is a symbol of failure and poverty. The scene conveys Soper's sense of devastation provoked by the realization that her son will be judged by his inability to live up to suburban standards of prestige. To draw upon another example, Martha Beck (1999, 190–191) says of her son Adam that "before he was even born, I was already comparing him to a disappointing holiday gift ... a gift everyone knows is inferior and broken, before it's even unwrapped." Beck draws on images of consumer culture and the rituals of gift-giving to describe her initial distress at the thought of having a child with Down syndrome. Rachel Adams writes about the "middle-class striving I had been taught to value" (Adams 2013, 17) and about "intellect and ambition, qualities my own background and life experience had taught me to value above all else" (36).

Many of these authors confess that a whole set of social expectations preceded their child's arrival. Kelly Anderson (2007, 103) writes that, before the birth of her son Will, "my life was indeed perfect. Three beautiful children, a caring husband, a nice house, a stylish nursery, and a new son on the way. The whole spotless picture. There was no room for flaws in my life." She evokes images of real estate and interior decorating to illustrate the standards of perfection she was striving to embody. For Andi Matthews (2007, 279), expectations about her family and her prospective son include dreams of milestone events in middle-class life: "I envisioned starting college funds, preparing for his wedding, having grandkids. I wanted the whole wonderful parenting package." On this understanding, parenting is seen as a kind of consumer product or package deal, and the success and joys of parenting are associated with seeing one's child realize a set of conventional goals such as going to college and getting married. It is striking how often the authors mention that, prior to the Down syndrome diagnosis, they looked forward to the time when the child would go off to college (e.g., Estreich 2011, 7; Soper 2009b, 258; Price 2007, 190; Sherman 2007, 172; Beck 1999, 253), as well as to their child's eventual wedding or marriage (e.g., Hampton 2012, 155; Kopp 2009, 115; Leach 2009, 94; Branam 2007, 209; Zeid 2007, 160).

Along similar lines, Emily Zeid (2007, 160) describes her life before the arrival of her daughter Emma: "We had the loveliest wedding.... I had everything perfectly planned. Shortly after the wedding we began to search for a home to call our own. We found the perfect house and sped through escrow—it was meant to be ... my pregnancy was uneventful. We underwent all the recommended prenatal screenings and diagnostic ultrasounds, and eagerly anticipated each doctor's appointment. Everything was normal.... I knew with certainty that she would be a genius. I had everything perfectly planned." In this story and others, the expected arrival of the child is part of a larger plan that includes, for instance, finding and decorating the perfect house (see also Peele 2009, 86–87; Anderson 2007, 103). In fact, the expectation that everything, including the baby, has to be "perfect," comes up often in these parent narratives (e.g., Adams 2013, 10; Hampton 2012, 3–4; Branam 2007, 209; Castellano-Farrell 2007, 237; Huffman 2007, 39; Beck 1999, 55).

Weddings, college graduation, buying a house are, of course, milestones whose attainment signify a successful middle-class life. But if the parents come to believe, because of the diagnosis of Down syndrome, that their child will not be able to reach these milestones, will not be "perfect" in this sense, everything they had expected goes awry. Though, in point of fact, as George Estreich (2011, 7) notes, some people with Down syndrome do go to college (Estreich 2011, 7), and many do marry.

Perhaps the most extreme example of the distance between middle-class aspiration and the assumed limitations of people with Down syndrome can be found in Martha Beck's *Expecting Adam*. Beck became pregnant with Adam while studying for her doctorate in sociology at Harvard. She describes the intense atmosphere of rationality, academic competitiveness, and achievement at Harvard as having been "my 'ideal' culture, since I was seventeen years old" (Beck 1999, 191). As a Harvard student, Beck stood at the apex of middle-class accomplishment. She goes on to say that, when she found out during her pregnancy that Adam had Down syndrome, "No one at Harvard seemed able to tolerate the thought of Santa's leaving the wrong kind of baby under the tree" (191).

This sense that the new child with Down syndrome will not be able to live up to middle-class expectations helps explain the parents' inchoate sense of loss that accompanies the diagnosis of Down syndrome in many of these stories. In a way, they see having a child with Down syndrome as a failure to realize certain middle-class ideals, such as the goal of perfection, or of witnessing one's child attain a conventional series of life milestones. When the child is diagnosed as having Down syndrome either before or after birth, these middle-class ideals control the narrative by, in effect, dictating to the parents and to those around them what to think and how to feel about this experience. Within the context of these ideals, it makes sense to think of the Down syndrome diagnosis as a loss of some kind. If the birth of such a child is considered at first to be a loss, it is because the child is perceived through a prism of a set of internalized middle-class values. To see the baby differently would then require a different set of values.

The Reorientation of Values

Of course, the story does not end there. All of these authors write with courage and honesty about their darkest and most unflattering emotions—feelings of bias, the urge to reject the baby, wanting to escape. But none of them would feel compelled to write these stories if their thoughts and emotions about their child didn't change—if they weren't able to write, as Estreich (2011, 20) puts it, "from the other side of acceptance." Many of the authors testify that they were personally transformed in a positive way by the experience of having a child with Down syndrome. The title of Kathryn Soper's book—*The Year My Son and I Were Born*—clearly implies that she was reborn through caring for her son. Martha Beck writes similarly of rebirth and transformation (Beck 1999, 7, 195). There are many other dramatic statements of personal change in this literature (e.g., Adams 2013, 193; Armstrong 2009, 21; Walker 2009, 337; Iannone 2007, 133).

Throughout these accounts of change, the authors often describe the new values they have welcomed into their lives. Many describe their embrace of a new sense of unconditional love toward their children (e.g., Hampton 2012, 13; Boggian 2007, 264; Groneberg 2007, 292). They also relate that values such as acceptance (e.g., Hampton 2012, 54; Foraker 2009, 38; Groneberg 2007, 292) and compassion (Hampton 2012, 99; Zeid 2007, 164; Bérubé 1996, 251) play a greater role in their lives. These narratives show that parents often undergo a kind of moral development after bringing a child with Down syndrome into their families. Sometimes this moral development takes the form of refiguring old values that they had previously endorsed, rather than adopting wholly new values. For example, a doctor as well as parent, Janis Gonzales writes of her changing understanding of perfection. She writes: "Medicine feeds the perfectionist beast inside us by insisting that it is possible—indeed, it is required—to never make mistakes.… Cariana's diagnosis of Down syndrome shattered my dreams of leading a 'perfect' life. But as

I got to know her, I realized that she wasn't flawed at all; what was flawed was my definition of perfection.... No test can tell us who our children will be, or what gifts each child might bring into the world." (Gonzales 2007, 249).[2] After first using the image of a fallen picket fence to illustrate her sense of loss, Kelle Hampton (2012, 54) revisits the image to describe her newfound understanding of perfection. She writes that, "My picket fence was being painted once again—a vibrant and rich hue.... I've come to *love* my fence, even though it may be different from the neighbors'. The concept of *perfection* is not flawless or ripped from a magazine. It's happiness. Happiness with all its messiness and *not-quite-thereness*" (Hampton 2012, 54 emphasis original). Hampton admits that although her previous desire for perfection set her up for grief when Nella was born, her understanding of perfection has since been transformed so that she can now see her child as perfect.

The child with Down syndrome is often named as the agent of change in bringing about this reorientation of values. In this role, the child is frequently described as a "teacher." Janine Steck Huffman (2007, 42) writes that her son Nash "has taught me a lesson I still work through every day—that the value of life, of a human, of a child, is measured not by how much he or she can accomplish, but how much he or she can teach others about what really matters. Like how to accept people with all kinds of different 'abilities'." Kelly Anderson (2007, 105) writes of her son Will that "I'd been blind to the wonder that surrounds us, the glorious, everyday miracles that most of us pay no heed to, the precious people we have no time for. How self-centered our whole existence can be—unless we have a teacher. Will is mine." Michael Bérubé (1996, 34) writes that, even though he does not subscribe to the myth that children with Down syndrome have an essential friendly and gentle nature, "I catch myself believing that people with Down syndrome are here for a specific purpose—perhaps to teach us patience, or humility, or compassion, or mere joy." These are just three of the many parent narratives that cast the child with Down syndrome in the role of teacher of what is truly important in life (see also, for example, Foraker 2009, 38; Schmidt 2009, 16; Crawford 2007, 127).

For Martha Beck, the lessons she learns through caring for her son Adam stand in sharp contrast with the values of Harvard University. Beck attributes her reorientation of values to the education provided by Adam. She says that, "I have had to unlearn virtually everything Harvard taught me about what is precious and what is garbage.... He has taught me to look at things in themselves, not at the value a brutal and often senseless world assigns to them" (Beck 1999, 317). Before learning Adam's diagnosis, Beck's husband, John, tells her why he is in favor of selective abortion: "'If a baby is born not being able to do what other people do, I think it is better not to prolong its suffering'" (qtd. in Beck 1999, 135). To which Beck responds by asking, "what is it that people *do*? What do we live to do, the way a horse lives to run?" (135). The exchange implies that her equally overachieving Harvard-educated husband believes that cognition is the essence of being human, and that cognitive disability makes a baby unable "to do what other people do." After Adam is born, on the other side of acceptance, Beck and her husband come to believe instead that, "This is the part of us that makes our brief, improbable little lives worth living: the ability to reach through our own isolation and find strength, and comfort, and warmth for and in each other. This is what human beings *do*. This is what we live for, the way horses live to run" (Beck 1999, 136). Beck's account in *Expecting Adam* documents her transition from endorsing the values supported by her Harvard culture to believing that one of the most important things that people do is care for one another.

In addition to the teacher metaphor, the parents writing these memoirs often use metaphors and images of shining, or of light, to describe their children. There are several examples in Jennifer Groneberg's *Road Map to Holland*, which focuses on the miniature and local moments of insight and mystery in her life while raising three sons, one of whom has Down syndrome. In one such moment, she finds a sparkly barrette with three stars in her garden. Later on, while shopping at a grocery store, Groneberg says, "I touch something in my hair and realize it's the sparkly barrette I'd found in the garden. Three stars, three sons, shining" (Groneberg 2008, 194). When her son Avery speaks his first words ("Ahluvyou!"), she comments that, "It's one moment in time, silver and perfect, shining and whole" (239). Similarly, Nancy Iannone (2007, 133) describes her daughter Gabriella as "shining brightly in our lives," and Alicia Culp (2007, 121) says of her daughter Kayla that "with the light that grows stronger in her every day, our path will always be brightly illuminated." These images of shining and light come up so often in these stories that Madeleine Will (2009, xvii) makes explicit mention of them in her foreword to the *Gifts 2* anthology: "One can't help but be struck by the use of images of light in the stories."

Of course, light is a common metaphor for knowledge, truth, goodness, and redemption in Western literature, philosophy, and religious thought.

Our cultural familiarity with this metaphor is one reason these authors reach for it in describing their children. But the images of shining and of light also tell us something important about how these parents come to see their children, and they stand in sharp contrast to the turmoil and sadness that the parents felt on first learning of the diagnosis of Down syndrome.

Theoretical Approaches

By presenting this rich genre of memoirs in the service of a larger argument in bioethics and reproductive ethics, I have moved into a disciplinary territory that overlaps with the medical humanities, and narrative medicine—fields that draw on literary and narrative sources to explore concepts such as health, illness, disability, and medical practice. I have taken these memoirs as presumptively reliable testimonies about the rewarding experience of parenting a child with Down syndrome. This evidence is presented in support of my position that more prospective parents ought to choose to bring such children into the world.

There are a number of theoretical approaches toward narratives like these parenting memoirs in the medical humanities, bioethics, philosophy, and literary theory that can be put to use in interpreting them. I have drawn upon two such approaches in my analysis of these memoirs. First, my position that social attitudes play a role in the reception of Down syndrome as a disability—as exemplified in some of these parent memoirs—has been influenced by the "social model" of disability. The "social model" holds that disability is caused not by the impairments that inhere in bodies or in brains, but instead by social arrangements that oppress people with these physical and cognitive differences. Second, my attempt to place the middle-class social context of these stories into the foreground contains traces of a Marxist critique of capitalism. I return to these two theoretical approaches—the social model of disability and the Marxist critique of capitalist influences on reproductive decision making—in chapters 7 and 8.

Aside from the two that appeal to me, there are many more possible theoretical approaches to Down syndrome parent memoirs. Below, I discuss several other theoretical approaches that seem to lend some insight into how meaning is created in the parenting memoirs under consideration here—into how the authors construct narrative meaning out of what they have experienced. This set of theories is not meant to be an exhaustive list of interpretations. Though some of the other approaches appear to be natural interpretive prisms through which we might view these memoirs, each has its limitations, some of which I have noted. The first of these other theoretical approaches that I explore returns to the prominent metaphors of light and shining that I have found in Down syndrome parenting memoirs.

Finding Meaning through the "Shining" Metaphor

In *All Things Shining*, philosophers Hubert Dreyfus and Sean Dorrance Kelly (2011) explore the significance of the "shining" metaphor in the Western tradition. Not at all about Down syndrome, their book instead describes how we can uncover a latent source of meaning in our demystified secular age. Dreyfus and Kelly (2011, 202) see the metaphor of shining as signifying the experience we have when the sense of ourselves, our world, and what matters to us comes into focus. We have this experience when something or someone "shines forth": for example, when we witness a feat of exemplary skill, heroism, or courage; or when we come in touch with something that is sacred that imbues our lives with meaning. In an earlier, polytheistic age, when people witnessed something shining forth, they would see this as a sign the gods were present and intervening in their lives. But even now, feelings of reverence and gratitude are appropriate attitudes when we meet with something or someone that shines forth.

With the frequent metaphors of shining or light parent memoirists use to describe their children with Down syndrome, they convey their sense of gratitude and reverence that these children are a part of their lives. For the authors in this genre, the experience of parenting a child with Down syndrome has transformed them and given their lives a meaning they had not known before. The details of personal transformation show this. In these stories, the parents move from an early stage of sadness or rejection to writing passionately about their personal growth through devotion and care for their children. In the process, they come to see that their initial attitudes of sadness and rejection were influenced by the priorities they had given to things that ultimately didn't matter.

Parental Virtues

Because the narratives of these parenting memoirs articulate values of unconditional love of their children, acceptance of difference, and compassion as parental virtues, virtue theory seems to be a natural theoretical lens through which to view their stories. There are broad similarities between what the memoirists and virtue theorists believe about good parenting, but also a few important differences that are worth mentioning.

Theorists of virtue ethics and others have presented values such as acceptance and unconditional love, or analogues to these values, as those of the virtuous parent (McDougall 2005, 2007; Sandel 2007; Strange and Chadwick 2010). According to Rosalind McDougall (2005), these values are virtues because they derive from parental attitudes that are conducive to the flourishing of children. For instance, acceptance is a parental virtue because children are born into extreme vulnerability and require a great deal of care for their survival and well-being. Parents must deal with the fact that the children they bring into the world will have unexpected qualities—technologies of prenatal selection cannot, and will not be able, in the final analysis, to allow us to control the characteristics and the fate of our children. Even total genetic information about our prospective children will not be able to account for the unavoidable complexity of gene-environment interactions that influence who our children will become (Strange and Chadwick 2010). To deal with the challenges of our children's vulnerability and unpredictability, while also providing the care needed for them to flourish, we must be motivated by a strong value of acceptance that prevents us from exposing our children to neglect and danger.

Jackie Leach Scully and colleagues (2006) conducted a study into moral decision making by lay research participants about reproductive ethics. The researchers found that, when asked to ethically evaluate genetic and reproductive technologies, participants tended to adopt a kind of virtue ethics framework emphasizing ideas about the characteristics of a good parent, rather than endorsing ideas about reproductive autonomy (Scully et al. 2006).[3] Within the virtue ethics framework, the participants endorsed values of relinquishing control over their children, and avoiding the impositions of expectations on children (Scully, Banks, and Shakespeare 2006)—attitudes that imply the need for acceptance of difference. These "lay" opinions about the attitudes of good parents are similar to the theoretical analysis of parental virtues provided by virtue theorists.

But the usefulness of virtue theory in assessing the parental values and attitudes expressed in Down syndrome parenting memoirs should not be overstated. As Stephen Holland (2011) points out, virtue theorists do not always arrive at the same conclusions about bioethical issues they explore.

If we understand the authors of these memoirs as themselves lay virtue theorists, their conclusions about the appropriate parental attitudes toward children and prospective children with Down syndrome might not fully cohere with the conclusions of other virtue theorists. Colin Farrelly (2007), for instance, finds that virtue theory would endorse prenatal genetic enhancement. Furthermore, even though Rosalind McDougall (2007) promotes "acceptingness" as a parental virtue, she takes a position disapproving of attempts by deaf prospective parents to prenatally select a genetically related child who would likewise be deaf. Neither Farrelly nor McDougall directly addresses the issue of selecting against Down syndrome, but their positions in favor of genetic enhancement and against embracing disability-related differences suggest that they might not agree that the parental values and attitudes documented in Down syndrome parent narratives are requirements of virtue.

I will return to similar themes about the parental values and attitudes most conducive to the flourishing of children in chapter 5. For now, it is important to note that, regardless of the differences between the memoirists and virtue theorists, Down syndrome parent memoirs clearly present a set of values that the authors believe embody the virtues of parents.

The Down Syndrome Parenting Memoir as Archetype or Rhetorical Strategy

The narrative of values transformation that I have found in Down syndrome parenting memoirs fits well into the archetype of "quest narrative" as described by Arthur Frank in *The Wounded Storyteller* (1995). Frank's book presents a taxonomy of illness narratives, one of which is the quest narrative. The "restitution narrative" and the "chaos narrative" are others.[4] According to Frank (1995, 115): "Quest stories meet suffering head on; they accept illness and seek to *use* it ... the quest is defined by the ill person's belief that something is to be gained through the experience." Quest narratives are stories of transformation. After undergoing transformation, at the end of the journey, the storyteller receives a "boon"—"the teller has been given something by the experience, usually some insight that must be passed on to others" (118). We can see this archetype reflected in Down syndrome parenting memoirs. Their narratives describe how parents and families become reoriented toward new values, and the lessons the authors have learned from their children.

The memoirists relay these messages, the "boon" of their transformation, back to their readers. An example of such a boon can be found in Soper's memoir. Soper relays a remark made by a friend, who says, "'It seems like parents of kids with disabilities know this big secret that the rest of us don't get to understand. It's like you're all in this club, and you have the key to life or something'" (qtd. in Soper 2009b, 295). The boon in the Down syndrome parenting quest narrative is the "secret lesson" about how to live, learned from the transformative experience of raising such a child. While the archetype fits fairly well, this formulation is perhaps overstated. The transformation experienced by raising a child with Down syndrome does not deliver "the key to life," and the lessons learned are no secret. What is more, unlike Frank's "quest narrative," the parenting memoirs are not illness narratives—the authors are themselves not ill, and, in most cases, neither are their children. However, the pattern of a "quest" as described by Frank is nonetheless recognizable in these narratives.

To describe and categorize Down syndrome parenting memoirs in this way might seem to cast doubt on their truthfulness. Because the memoirs follow a recognizable pattern, one might suspect that the real-life parenting experiences the memoirs are meant to document might have been wrestled into a standard story arc for the purposes of meeting conventional expectations about the meanings that such stories should convey. These memoirs might not be mirror-like representation of actual events exactly as they occurred, as they are rhetorical inventions or constructions. Some have argued that all memoirs are rhetorical constructions whose purposes go beyond relating events and emotions as they actually happened. For example, G. Thomas Couser's work on disability memoir, *Signifying Bodies: Disability in Contemporary Life Writing* (2009) argues that there are a number of rhetorical schemas that are available to authors of memoirs. Couser evaluates disability memoirs according to whether the chosen rhetorical schema among several available to memoirists supports a politically progressive understanding of disability. His account emphasizes the constructed aspect of memoir, as a genre.

However, in response to this line of argument, even if a given memoir is a rhetorical construction, this does not immediately force us to question its truthfulness or the extent to which it represents either a memoirist's experiences and emotions as they occurred during the time period recounted or the memoirist's belief system. A memoir can be a truthful representation of events while at the same time conforming to one or more rhetorical schemas. Furthermore, the meanings that are derived from Down syndrome parenting memoirs are primarily about the transformed values the parents have come to endorse. In the foregoing sections of this chapter, I have related these stories of value transformation. These claims—since they are reporting on their own internal or "psychological" reactions—are not the kind that can be criticized for failing to accurately represent events as they occurred. The parent memoirists make intimate subjective assessments of their beliefs and values, as they have arisen out of the experiences of their families in raising children with Down syndrome. These subjective attempts to make meaning out of their own experiences cannot be impugned on epistemological grounds (especially by arguments that would so impugn *all* memoirs) even if the memoirists' narratives happen to conform to well-worn rhetorical schemas.

Narrative Ethics

Narratives, such as the ones written by these parent memoirists, can provide meaning to the events of our lives. Without a sense of connection or unity, a person's life can seem like a series of incoherent, unrelated experiences. Stories that we tell about ourselves and our loved ones can provide the missing unity and create meaning out of our life events. In Down syndrome parenting memoirs, we witness these authors struggling to provide coherence to two sets of experiences. In the first phase after the births of their children, they tell a story of sadness and grief. After they come to embrace their children and to reorient their values, they enter a second phase and tell a second story, one of acceptance and moral development. The first story corresponds to a common cultural narrative in which the birth of a child with Down syndrome is something to be feared because of the hardship and disappointment associated with caring for a child with a disability. The grief these authors experience is influenced by this common cultural narrative. According to narrative ethics theorists such as Howard Brody and Hilde Lindemann Nelson, moral agents are able to create counternarratives about their lives when the events they experience and the stories that others tell about their lives disrupt, distress, or damage their identities (Brody 1994; Nelson 2001; see also Chambers 2010). The second story told by these memoirists constitutes an effective counternarrative—re-creating unity and meaning in their lives that is worthy of what they experience.

Importantly, these counternarratives also fulfill an ethical obligation that we bear toward people who are living with Down syndrome. The common cultural story about the birth of a child with Down syndrome casts people living with this condition as agents of turmoil and unhappiness, people whose very birth is an occasion for grief. This cultural perception of people with Down syndrome cannot be positive for their well-being. By constructing a counternarrative, the authors we have examined in this chapter resist this unfair depiction of people who have Down syndrome. In their stories, the child with Down syndrome is recast as an agent of happiness. We have an ethical obligation to prevent the spread of false and potentially harmful cultural narratives about vulnerable groups of people, and Down syndrome parenting memoirs provide this ethical service.

Possible Objections to My Use of Parenting Memoirs

The parenting memoirs I have presented here recognizably constitute a convenience sample that is prone to bias. The parent memoirists are a self-selected group predisposed to report on their overall positive experiences of parenting a child with Down syndrome. As such, their memoirs cannot in themselves be taken as representative of all the experiences of parenting a child with Down syndrome. But the findings of empirical social science research into the lives of parents and families with a child who has Down syndrome generally support their claims. That is, evidence from the social sciences supports the idea that parenting a child with Down syndrome can bring about positive effects within families and be a source of personal growth. A recent study by Brian G. Skotko, Susan P. Levine, and Richard Goldstein shows that an overwhelming majority of parents of children with Down syndrome love their children, have pride in them, and experience joy and rewards from raising a child with Down syndrome (Skotko et al. 2011b). Furthermore, such parents tend to "cite life lessons in acceptance, patience, and purpose" (Skotko, Levine, and Goldstein 2011b, 2344). The research by Skotko, Levine, and Goldstein was based on survey responses from over 2000 parents of children with Down syndrome in the United States.

A second study by Gillian King and colleagues (2006), which used a qualitative focus group methodology, reached conclusions about parenting a child with Down syndrome that are strikingly similar to the accounts of the parenting memoirs. The parents in the focus groups (which included parents of children with autism as well) reported that, before the birth of their disabled child, they placed undue emphasis on values such as accomplishment, social recognition, and independence (King et al. 2006, 358). As one participant put it, she had "dreams as a mother for this child, dreams of academic achievement, dreams of marriage, dreams of children" (358). But the birth of the child brought about a reexamination of beliefs, and a new appreciation of values such as "patience, acceptance, tolerance, perseverance, compassion, and unconditional love" (361). The studies by Skotko, Levine, and Goldstein and by King and colleagues are supported by other research into the functioning of families with children who have Down syndrome—research that uses a variety of methodologies and that studies several different indicators of good family functioning (e.g., Cunningham 1996; Cuskelly, Chant, and Hayes 1998; Cuskelly and Gunn 2006; Duis, Summers, and Summers1997; Dumas et al. 1991; Kaminsky and Dewey 2002; Noh et al. 1989; Rodrigue, Morgan, and Geffken 1992, 1993; Stores et al. 1998; Thomas and Olson 1993; Urbano and Hodapp 2007; Van Riper, Ryff, and Pridham 1992; Van Riper 2000). In chapter 4, I will discuss this large body of consistent and compelling social science evidence that parenting a child with Down syndrome does not cause dysfunction within one's family, and that introducing a child with Down syndrome into a family has about the same effect on family functioning as introducing a child without a disability.

A second objection to the use of parenting memoirs as a depiction of what it is like to parent a child with Down syndrome has been raised by Alison Piepmeier (2012) in her article "Saints, Sages, and Victims: Endorsement of and Resistance to Cultural Stereotypes in Memoirs by Parents of Children with Disabilities." Piepmeier criticizes a number of the memoirs we have examined here for dwelling too much on the parents' feelings of grief. She suggests that when the authors of these memoirs depict the grief they experience upon learning the diagnosis of Down syndrome, they are supporting the cultural stereotype of "children with disabilities as damaging forces in their parents' lives" (Piepmeier 2012). However, I would argue that it would be an odd affront to honesty to leave these experiences out of the narrative. Piepmeier concedes that "grief serves a dramatic function" in the lead-up to depicting the transformation of these parents. There would be no transformation without a starting point subsumed by emotional distress. She also agrees that the honesty in these memoirs is "an important validation" (Piepmeier 2012) for parents undergoing similar experiences. Piepmeier's problem with expressions of grief is a matter of emphasis. She says "many of the memoirs focus so fully and in such great detail on the grief that this is the overriding emotion of the books" and that "because the turning point often happens near the end of the book, then the majority of the reading experience has consisted of those false stereotypes being bolstered" (Piepmeier 2012). She lauds memoirs such as Michael Bérubé's, which devotes as few pages as possible to accounting for the emotional turmoil he experienced when his child was diagnosed with Down syndrome (Piepmeier 2012).

I can only say that Piepmeier's reading of these works is idiosyncratic. As I have demonstrated above, the lasting impression for most who read any of these memoirs is one of affirmation that parenting a child with Down syndrome is a rewarding and beneficial experience. The emotional depictions of guilt and sadness that mark the beginning of most of these memoirs are overshadowed and put into question in the end by the authors' descriptions of the humanity and value of their children.

A third objection to the use of parenting memoirs in support of my position is that the insight and personal development attributed by these authors to parenting a child with Down syndrome may not be unique to Down syndrome. Similar kinds of insights and changes can be brought about by parenting children with other disabilities. I would agree with this point. If we widen our lens beyond Down syndrome, books written by and about parents of children with other disabilities have clear similarities with the memoirs examined here. For example, *This Lovely Life* by Vicki Forman (2009), *The Boy in the Moon* by Ian Brown (2009), *Love's Labor* by Eva Feder Kittay (1999), and *Far from the Tree* by Andrew Solomon (2012) all relate that parents of children with even severe disabilities experience joy by having these children in their lives, and that such parents often undergo a form of moral development similar to that described by parents of children with Down syndrome.

My primary goal in this chapter has been to argue that more prospective parents should choose children with Down syndrome rather than selectively terminating affected pregnancies, because as parenting memoirs show, having a child with Down syndrome can be highly rewarding and personally transformative. In the next three chapters I will address, in detail, three more substantive objections, one to this argument and two others to my larger position.

Conclusion: Parenting Memoirs and Selective Termination

One aspect of the discourse about parenting children with Down syndrome that is not shared with the discourse about parenting children with certain other disabilities is the availability of early prenatal testing and selective termination for this condition, though, of course, this is changing with the rapid expansion of such testing. As I pointed out earlier, the selective termination rates or overall numbers of selective terminations for Down syndrome, or both, are likely to increase as a result of the growing availability of noninvasive early prenatal testing. When one reads about the crises that parents of children with Down syndrome go through when given the diagnosis, it is not surprising to learn that 60 percent or more of prospective parents elect to terminate when they are given this information prenatally. But the attitudes toward Down syndrome that motivate the decision to terminate stand in stark contrast to the attitudes that the authors of Down syndrome parenting memoirs eventually come to have toward their children. These authors and their families lead flourishing lives. This contrast suggests that many selective termination decisions are motivated by a lack of understanding about Down syndrome. The stated goal of many of the parenting memoirs and much of the social science research I have cited is to foster a greater public understanding of Down syndrome that might help new and prospective parents who are faced with a Down syndrome diagnosis.

The parenting memoirs we have examined reveal that cultural attitudes play a significant role in the constitution of a disability. So we should recognize that high selective termination rates for Down syndrome are due, to a large extent, to the values prevalent in our culture—what we think is valuable, and what we devalue—rather than being due to any inherent deficit or deficiency of people who have Down syndrome.[5] If we lived in another culture that valued different things, we would not regard the diagnosis or birth of a child with Down syndrome, even initially, as a loss. In our own culture, in accordance with the better angels of our nature, we already claim that we esteem many of these "different things": unconditional love for our children, compassion, acceptance of difference. Through causing the reader to reflect on these values, these memoirs written by parents of children with Down syndrome not only present a different way of thinking about cognitive disability; they also force us to think about what kind of society we want to live in.

4 Adaptive Preference and Empirical Research about Families of Children with Down Syndrome

The memoirs written by parents of children with Down syndrome provide powerful testimony that bringing a child with Down syndrome into a family typically benefits the family and may contribute to its flourishing. The point of offering such testimony is to suggest that prospective parents should seriously consider having a child with Down syndrome at the various decision points in the process of prenatal testing and when offered the option of selective termination if given a prenatal diagnosis. The decision to have a child with Down syndrome would go against the societal trend of selective termination for this disability.

There are a number of possible responses to the argument that prospective parents should seriously consider having a child with Down syndrome because of the positive assessments provided by parents of such children. First, some might question whether the testimony of parents of children with Down syndrome is reliable—whether their descriptions of family life are distorted by "adaptive preference," which emphasizes positive views, but overlooks dysfunction. Second, some might argue that it is presumptively wrong to bring a child with a disability into the world, for the child's own sake, when you could otherwise try to bring a nondisabled child into the world. A third possible response would be that, even if it is not presumptively wrong to bring a child with a disability into the world, there is still nothing wrong with selective termination in order to try for a nondisabled child. According to this third response, because no one is harmed by selective termination, starting over in order to have a nondisabled child is a morally innocuous choice. In the next three chapters I will address these three responses, and show the deficiencies with each.

Response 1 (chapter 4): Down syndrome parenting memoirs are affected by self-deception or adaptive preference and are therefore unreliable;

Response 2 (chapter 5): It is morally wrong to bring a child with Down syndrome into the world when you could try for a child without a disability;

Response 3 (chapter 6): There is nothing wrong with selective termination of a pregnancy to prevent the birth of a child with Down syndrome in order to try for a child without a disability.

These three responses to the argument in favor of having a child with Down syndrome are each disputable. For instance, it is particularly difficult to prove parental self-deception or adaptive preference (response 1) across all of the sources I have cited. Furthermore, the burden of proof is particularly high when alleging that someone has committed a moral violation (response 2) by continuing a pregnancy after being given a prenatal diagnosis of Down syndrome. The position advanced in response 3 is perhaps the least contentious and consequently the most difficult to counter—but I will do so in chapter 6. In this chapter, I begin with the first response: the idea that parenting memoirs describing family life with a child who has Down syndrome may be affected by self-deception or adaptive preference.

Adaptive preference is a common psychological mechanism among people who live in difficult circumstances—oppression, poverty, resource scarcity—in which they state that they are satisfied with their circumstances. Keeping an optimistic outlook can be a coping mechanism when faced with dismal prospects. For instance, in the Charles Dickens novel *A Christmas Carol,* Tiny Tim is a happy and generally optimistic character, despite being impoverished and living with a disability. Readers quickly come to understand, however, that Tim's happy and optimistic behavior is not a reliable guide to his actual well-being (Wasserman, Bickenbach, and Wachbroit 2005, 11). Adopting a positive outlook can be a way of adapting to deprived circumstances. Those engaging in such adaptive behavior might even claim to prefer living in such circumstances, or to prefer not to change their circumstances if given the opportunity—hence "adaptive *preference.*"

The memoirs written by parents of children with Down syndrome are, in essence, detailed subjective statements of personal satisfaction with their family lives. Because of the subjective nature of these assessments and the possibility of adaptive preference, it is only reasonable to ask whether the statements by parents of children with Down syndrome about the positive effects on their families of having children with Down syndrome are in fact reliable indicators of the well-being of such families. If other evidence indicates that these families actually don't tend to function well, or that the parents are somehow deceiving themselves, this would undermine the case for encouraging future parents to have children with Down syndrome.

Underlying this line of response is the view that the well-being of an individual or a family has both a subjective and an objective aspect (Brock 1993). Subjective assessments of satisfaction are certainly relevant to an overall judgment of whether a family is doing well. But, according to this view, these subjective assessments are not fully authoritative. Subjective testimony about a situation can put an overly positive (or negative) "spin" on assessments of that situation. In order to arrive at a well-supported judgment of well-being, it might be necessary to look at more objective indicators. In the case of Tiny Tim, one wonders whether he has enough food, or whether he has access to health care that he needs. In the case of families with children who have Down syndrome, objective indicators of family well-being would include divorce rates, observational studies of family function, and behavioral profiles of siblings of children with Down syndrome. This chapter will investigate these indicators.

Reasons for Alleging Adaptive Preference

Despite the extensive testimonials found in parenting memoirs about good family functioning in families who have a child with Down syndrome, the assumption that having a child with a disability imperils family life persists in the minds of many people. For example, in their influential book *From Chance to Choice,* Allen Buchanan and colleagues (2000, 275) mention the "avoidable and serious strains on [a] marriage or family" caused by having a child with a disability that would justify avoiding the birth of such a child. In his book *Far From the Tree,* Andrew Solomon (2012, 191) discusses parental affirmations of having a child with Down syndrome, and questions whether these affirmations are a false construction by parents "to disguise their despair." On this issue Solomon interviewed parents, activists, and psychologists. He says, "I met disability activists who insisted that everyone's joy was authentic, and I met psychologists who thought no one's experience was" (191). Though Solomon himself endorses neither position, it is clear that many professionals and lay people disbelieve the claim that having a child with Down syndrome is typically a positive experience. Furthermore, there is a literature that states that having a child with Down syndrome in one's family is almost always harmful to the members of the family. These are arguments that could be taken to support an allegation of adaptive preference and the view that having a child with Down syndrome threatens a family's basic flourishing.

Kuhse and Singer—Should the Baby Live?

Examples of the view that a child with Down syndrome is destructive of family life can be found in the book *Should the Baby Live? The Problem of Handicapped Infants* (1985) by Helga Kuhse and Peter Singer. Kuhse and Singer argue that parents should be allowed to kill infants born with Down syndrome or any of several other disabilities.[1] The word they actually use is "euthanize," but I take issue with the idea that Down syndrome is a condition for which the concept of euthanasia even makes sense, given that killing in the sense of "euthanasia" would only be appropriate if the patient were experiencing extreme pain and suffering that could be neither cured nor substantially alleviated. One of the reasons Kuhse and Singer give for their position is that looking after a child with Down syndrome is invariably a "strain" on families (Kuhse and Singer 1985, 147), and quite often "destructive of family harmony" (152). They cite studies from the 1970s and 1980s of "objectively measurable features of the lives of families with handicapped children" (149). These "objective" measures of family strain might be taken to support a claim that subjective reports of family harmony are a product of adaptive preference, or that families with a child who has Down syndrome are in danger of failing to flourish (149). The studies from the 1970s and 1980s also suggest that "these families have a higher rate of marital break-up"(149), that, even when the child with Down syndrome has a mild disability, there is a 50 percent chance that one of the child's siblings will be "disturbed" (150), and that parents of children with Down syndrome are almost always tired and depressed, unable to have a social life, and unable to leave the house for daily activities (148). The overall impression created by Kuhse and Singer is that having a child with Down syndrome causes family dysfunction. They readily admit to the existence of data that show many such families are satisfied with the lives, but they also tend to discount this evidence—dismissing as "optimistic," for example, the claim by Cliff Cunningham (1982) that up to 75 percent of families who had a child with Down syndrome saw the experience as rewarding (Kuhse and Singer 1985, 152).

In addition to objective studies, much of the evidence provided by Kuhse and Singer is subjective anecdote—usually depicting dysfunction.

They counter the evidence of positive experiences of raising a child with Down syndrome with stories in which families are "totally wrecked" (Kuhse and Singer 1985, 152) by (they say) the presence of a child with cognitive disabilities. In order to reinforce their contention that a child with Down syndrome usually causes dysfunction, they cite the experiences of Charles Hannam as "a median point between the extremes" (Kuhse and Singer 1985, 152) of positive experiences on the one hand, and "wrecked" families, on the other (152).

Hannam, the father of a child with Down syndrome, was the author of *Parents and Mentally Handicapped Children* (1980). Rather than being a voice of moderation, as Kuhse and Singer claim, Hannam's position is actually quite extreme itself. Kuhse and Singer quote at length from Hannam's book in which he opens up about his feelings toward his son David. The passage includes the following admissions: "I am so sorry, David, I don't want you. I have tried to be good but I haven't done very much for you.... I have thought about murdering you and you are my child.... I have felt that you were my own special little albatross, hanging around my neck forever" (Hannam 1980, 46). Oddly Kuhse and Singer (1985, 147) refer to Hannam's book, in which this passage appears, as a "guide to other parents with Down's syndrome children." I cannot imagine how statements like these could be helpful in any way to parents of children with Down syndrome. Kuhse and Singer follow up Hannam's quotation by pointing out that "Shortly afterwards, the Hannams put David in an institution" (Kuhse and Singer 1985, 153).

If the depiction of raising a child with Down syndrome provided by Kuhse and Singer is accurate to any degree, then there is reason to doubt the reports of positive experiences provided by parents of children with Down syndrome in memoirs and in social science research. If Kuhse and Singer are right, there might be some grounds for an allegation of adaptive preference. However, as I will show, the experiences depicted in Kuhse and Singer (1985) are simply not representative of what it is like to raise a child with Down syndrome.

Conceptual Difficulties with Alleging Adaptive Preference

Though the response of suspecting that parenting memoirs are affected by adaptive preference is the weakest of the three possible responses I outlined at the beginning of the chapter, we ought to take it seriously. If only because of the challenges posed by Kuhse and Singer's work, and common opinions about raising a child with Down syndrome, it is necessary to counter the suggestion of adaptive preference. I will begin by noting some reasons to be hesitant about alleging that the assessments by parents of children with Down syndrome are affected by adaptive preference. In the parenting memoirs discussed in chapter 3, there is no reason to believe that the authors are self-deceptive in any way about the functioning of their families. In fact, the opposite is the case: the authors show themselves to be deeply reflective and painfully honest. Furthermore, to allege that parents of children with Down syndrome are deceiving themselves when they say that they and their children are highly fulfilled is actually quite demeaning and patronizing.

There is also a conceptual difficulty associated with viewing these parent narratives through the lens of adaptive preference. According to one well-developed philosophical account—Serene Khader's deliberative perfectionist approach—adaptive preference can be defined as "preferences inconsistent with basic flourishing that a person developed under conditions nonconducive to basic flourishing and that we expect her to change under conditions conducive to basic flourishing" (Khader 2011, 17). In Khader's account, objective aspects of a person's life are at odds with the person's overly optimistic subjective assessment and this combination leads to adaptive preference. According to Khader, the person's life is objectively going poorly because of a lack of basic flourishing. For Tiny Tim, his sunny outlook stands in stark contrast to the poverty and deprivation in which his family lives. His life lacks basic flourishing. The key concept here that creates the difficulty with viewing Down syndrome parent narratives as distorted by adaptive preference is the concept of "basic flourishing." A person's life is going well in the sense of "flourishing" if "she is exercising certain valuable capacities that it is in the nature of human beings to exercise" (Khader 2011, 49). There is a large body of literature on what these valuable capacities are, as well as discussions about the different ways such capacities can be exercised in a human life (see, for example, Nussbaum and Sen 1993). Capacities such as our ability to create social relationships, use our cognitive abilities, and make choices about our sexual lives usually make it onto the list of valuable capacities necessary for flourishing.

Because of the huge degree of human cultural variability, there is also huge variability in what it means to lead a good human life. There are many different ways to flourish. Nonetheless, there are also many ways in which human lives can go badly and fail to flourish. But in order to fail to achieve basic flourishing, we must usually experience serious deprivation—extreme poverty, oppression, normalized violence, sanitary conditions not conducive to health, for example. Clearly, it would be difficult to argue that parents raising children with Down syndrome who have written memoirs, or who are interviewed by social scientists, experience these kinds of deprivation. There is little reason to believe that having a child with Down syndrome would threaten basic flourishing in the way that serious deprivation would. In the analysis of adaptive preference, it is important to be aware that people and families can flourish in different ways. Khader (2011, 12) argues that one of the occupational hazards of alleging adaptive preference is "confusing difference with deprivation ... treating unfamiliar preferences that are fully compatible with flourishing as though they were adaptive preferences." A family with a child who has Down syndrome may be different from families who have only typically developing children. But this difference does not indicate deprivation on a scale that would impede basic flourishing. As chapter 3 shows, we can expect that parents of children with Down syndrome will experience adversity. These parents tend to go through a period of adjustment and coping, but the ability to cope with a diagnosis of Down syndrome is not indicative of a distorting effect on their assessment of how well their families are functioning. Khader's account suggests that when one lives in circumstances conducive to basic flourishing, being able to cope with adversity is actually a constituent of good quality of life. It is only when basic flourishing is absent that we may legitimately suspect the distorting effects of adaptive preference.

For these reasons, it is a weak objection to allege adaptive preference when presented with Down syndrome parenting narratives as an argument in support of having a child with Down syndrome. As I will show in the next section, social science evidence also underlines the weakness of the claim that the parenting memoirs are examples of adaptive preference.

Findings in Support of Family Flourishing

As I mentioned in chapter 3, in the survey of parents of children with Down syndrome conducted by Brian G. Skotko, Susan P. Levine, and Richard Goldstein (2011b), 99 percent of 2,044 respondents said that they loved their children who have Down syndrome and 97 percent reported feeling proud of them. Only 11 percent reported that their children with Down syndrome put a strain on their marriage—the same percentage that reported that their other children put a strain on their marriage. Only 4 percent reported any regret about having a child with Down syndrome (Skotko, Levine, and Goldstein 2011b). The same research team conducted surveys of siblings of children with Down syndrome (822 participants), and of 284 people with Down syndrome 12 years of age and older (Skotko, Levine, and Goldstein 2011a, 2011c): 96 percent of the siblings reported feeling affection toward their brother or sister with Down syndrome; 94 percent reported feeling pride (Skotko, Levine, and Goldstein 2011a). Of the people with Down syndrome themselves, 99 percent reported that they were happy with their lives, and that they loved their families (Skotko, Levine, and Goldstein 2011c). These robust findings are much more up to date and convincing than the data provided by Kuhse and Singer (1985). It is difficult to believe that such high percentages of people occupying different roles in the family structure could be deceiving themselves about the quality of the relationships within their families.

Quite a lot of research into the effects of introducing a child with Down syndrome into a family has been done since Kuhse and Singer's 1985 book, and virtually none of it supports their depiction of infants with Down syndrome as agents of dysfunction in their families. As a significant counter to the allegation of adaptive preference, some of this research uses objective measures of family well-being. Divorce rates are an example. In one longitudinal study in the United Kingdom, the divorce rate of families with a child with Down syndrome was lower than the national average (Cunningham 1996). A more recent population-based study in the United States comparing a cohort of families with a child with Down syndrome to families with nondisabled children found the divorce rate to be lower in the Down syndrome family group (Urbano and Hodapp 2007). A study by Volker Thomas and David H. Olson in 1993 used observational assessments to measure family cohesion, adaptability, and communication among four groups of families with children who had different behavioral issues (Thomas and Olson 1993). In this study, the families with children who had Down syndrome were considered a control group since their observed measures of family cohesion, adaptability and communication were virtually identical to a second control group of families with developmentally normal children who had no behavioral problems (Thomas and Olson 1993).

In addition to the research using objective measures, quite a few studies have used validated self-report measures—such as surveys, questionnaires, and inventories—to gain an understanding of families having children with Down syndrome. Such studies often compare families that include a child with Down syndrome with control groups of families who have only typically developing children. In such studies, levels of marital satisfaction are similar between the Down syndrome family group and the control group (e.g., Noh et al. 1989; Rodrigue, Morgan, and Geffken 1992; Van Riper, Ryff, and Pridham 1992). Parenting stress levels are also similar between the two groups (e.g., Dumas et al. 1991; Duis, Summers, and Summers 1997; Stores et al. 1998). There is no difference in the level of behavioral problems among the siblings of children with Down syndrome as compared to families with typically developing children (Rodrigue, Morgan, and Geffken 1993; Cuskelly, Chant, and Hayes 1998). Siblings of children with Down syndrome and siblings of typically developing children also show similar levels of social competence (Rodrigue, Morgan, and Geffken 1993; Van Riper 2000; Kaminsky and Dewey 2002; Cuskelly and Gunn 2006). The studies using self-report measures provide strong corroboration of the testimony found in parenting memoirs and in the research undertaken by Skotko, Levine, and Goldstein (2011a, 2011b, 2011c).

According to this body of research, the difficulties of parenting a child with Down syndrome are not as dramatic as Kuhse and Singer (1985)—and the research they cite from the 1970s and the early 1980s—would lead their readers to believe. However, the more recent research notes a few challenges faced by these families. There is empirical evidence for an increase in parental stress when the child with Down syndrome has a higher level of behavioral problems (Noh et al. 1989; Stores et al. 1998). Though children with Down syndrome tend to have a higher incidence of behavioral problems than typically developing children (Stores et al. 1998), the association with increased parental stress is probably true of all children who display difficult behavior. Some studies support the idea that parents of children with Down syndrome have lower levels of well-being than parents of typical children, possibly due to psychological effects related to managing behavioral difficulties, though other studies contradict this finding (see, for example, Cuskelly, Hauser-Cram, and Van Riper 2008).

One study (Brown et al. 2006) also documents lower levels of satisfaction about several aspects of family life among a group of families who have a child with Down syndrome, as compared to a control group of families whose children are all typically developing. The Down syndrome group had lower levels of satisfaction in the areas of health, financial well-being, support from other people, and careers (Brown et al. 2006).[2] However, respondents for the Down syndrome group expressed the same level of satisfaction with their family relations (89%) as those for the control group (88%; Brown et al. 2006).

Explaining the Conflicting Evidence

There is a clear contrast between the depictions of parenting a child with Down syndrome presented by Helga Kuhse and Peter Singer (1985) and those presented by recent empirical research. Kuhse and Singer present an inaccurate picture. Part of the reason for this contrast lies in Kuhse and Singer's rhetorical efforts to emphasize stories of families falling apart in order to make the counterintuitive case that parents should be allowed to kill their disabled infants. Because of the shocking nature of their position, they need to provide especially strong evidence to support it. Kuhse and Singer must present their evidence in a way that makes parenting a child with Down syndrome seem like an invariably harmful experience for all involved. Presenting the evidence for this purpose would mean, for instance, dismissing, downplaying, or ignoring evidence to the contrary. I mention an example of this strategy earlier in the chapter: Kuhse and Singer (1985, 152) characterize an empirical study documenting the rewarding experience of parenting a child with Down syndrome as "optimistic."

As a technical matter, it is more difficult to use a philosophical argument to defend a deeply counterintuitive position—such as arguing in favor of the ethics of killing infants with Down syndrome—than it is to argue for a more commonly accepted position. Viewed simply from the perspective of evaluating their use of argumentation to support a position many would regard as indefensible, Kuhse and Singer are highly successful. However, the difficulty of the task may have diminished motivation to explore contrary evidence or alternative explanations for the apparently negative experiences of parenting a child with Down syndrome. Any attempt to do so would have diminished the strength of their message that being such a parent is invariably a devastating experience.

A further explanation for the divergence between Kuhse and Singer's account and the current empirical evidence lies in what philosophers value and what they regard as lacking value. Some authors believe that it is almost inevitable that parenting a child with Down syndrome will be seen as tragic by philosophers. The philosopher Eva Feder Kittay (2005, 107), for example, suggests that "Perhaps philosophers are merely self-absorbed—even narcissistic. We have a tendency to valorise that which *we* do—above all things. Maybe that is why the exercise of reason gets elevated to that which is the essence of what it is to be human in philosophical writings." A child born with a cognitive disability will not be able to exercise the reasoning skills of a philosopher. Bringing such a child into the world could easily be seen as tragic by someone who values rationality above all else. Licia Carlson makes this point as well: calling intellectual disability "a philosopher's worst nightmare" (Carlson 2010, 4). Carlson compares philosophers' depiction of intellectual disability to Hollywood's portrayal of blind characters—who are portrayed as weak, pitiable, or disagreeable. (Carlson references Georgina Kleege's 1999 book *Sight Unseen*). The filmmaker is disturbed by the blind person, because such a person can't be a viewer—can't appreciate the filmmaker's art. Consequently, on the silver screen the blind person becomes someone of diminished standing, or someone lacking in social value. Similarly, for the philosopher, the infant with Down syndrome is depicted as a parental nightmare, as someone who can justifiably be killed. The child with Down syndrome is someone who can't appreciate the genius of the philosopher, and thus becomes the object of the philosopher's resentment.

The contrast between Kuhse and Singer's account and current research can also be explained through an analysis of changes in the methods and presuppositions of family functioning research over time. There has been a theoretical reorientation in studies of families with children who have disabilities in the years since publication of *Should the Baby Live?* in 1985. This reorientation is perhaps the main reason for the contrast between current research and the depiction of families with a child who has Down syndrome in *Should the Baby Live?* Cliff C. Cunningham (1996, 88) explains: "Early research on families of children with disability reflected a pathological model in that families were automatically assumed to suffer as a consequence of the child—in effect the studies only looked for negative outcomes." Ronald Gallimore and colleagues (1989, 216) make reference to a methodological turn from "pathology theories to ecological conceptions of family adaptation."

With improved research methods in the late 1980s and into the 1990s, a different picture emerged that was altogether different from the pathological model's picture. One aspect of this improvement was the willingness to accept that a difficulty experienced by a family with a child who has a disability might be due to a lack of support services, rather than to the disabled child. For example, some of the anecdotes cited by Kuhse and Singer (1985) are purportedly examples of the negative impact of the presence of a child with a disability on the parents' social life. One couple they quote attributes their poor social life to the fact that they can't find a reliable babysitter for their child who has Down syndrome (Kuhse and Singer 1985, 148). Kuhse and Singer repeat several times that the burden of caring for a child with a disability usually "falls on the mothers" (148). They argue that the child with Down syndrome is the cause of these problems, but they also assume that social circumstances must remain static. With help from social programs that provide qualified supplementary child care, the couple with babysitting problems could have had a satisfactory social life. In a social setting in which there is a more equitable distribution of child care responsibilities within families, along with support from other caregivers, the burden of child care would not be borne entirely by the mother.[3] Instead of these quite reasonable and practicable solutions, Kuhse and Singer recommend the option of killing the infant. More recent research incorporates variables such as access to social services in efforts to study the well-being of families (Skotko, Levine, and Goldstein 2011b).

The difference between the picture presented by Kuhse and Singer (1985) and the picture presented by recent empirical research may also reflect the transition from a situation where people with Down syndrome were subject to widespread institutionalization to one where people with Down syndrome primarily live with their families. This transition was still playing out in the 1970s—the decade in which much of the research cited by Kuhse and Singer was conducted. The process of deinstitutionalization occurred very rapidly. In 1967, the total number of people with cognitive disabilities who were housed in institutions in the United States reached a peak of 194,650 (Metzel 2004, 432). At the same time, alternative social programs and living arrangements within communities were being developed, often by parents of children with cognitive disabilities who chose to raise their children at home. Some parent groups started their own schools and educational programs when their children with cognitive disabilities were denied admission to public schools. The creation of the New York Association for Retarded Children was a consequence of this kind of initiative (Rothman and Rothman 2004). In the late 1960s and early 1970s, large public institutions and community programs operated in parallel and often competed for public funding (Metzel 2004). But the collapse of the institutional system, which occurred largely as a result of media attention and litigation, changed everything.

In 1972, a young Geraldo Rivera broadcast an exposé of Staten Island's Willowbrook State School on a New York City television station. Rivera presented shocking footage of children exposed to disease and violence, living in squalor with very little supervision, education, or health care, and very little chance to play as children (Rothman and Rothman 2004). The class-action lawsuit against the state of New York mounted by the New York Civil Liberties Union led to the deinstitutionalization of most of Willowbrook's residents throughout the 1970s and the closure of the institution in 1987 (Rothman and Rothman 2004, 445–446). Other major U.S. lawsuits supporting deinstitutionalization were also brought in the 1970s—*Wyatt v. Stickney* in 1972, and *Halderman v. Pennhurst State School and Hospital* in 1977 (Metzel 2004, 434). These lawsuits documented the horrors experienced by these most vulnerable children. The Partlow State School and Hospital in Alabama, subject of *Wyatt v. Stickney*, had "its floors and walls covered with urine and excrement" (Rothman and Rothman 2004, 457). In Willowbrook, "The gross disabilities and bizarre behavior that the visitor saw, limbs twisted into brambles and residents banging their heads against the wall, were not the reason for incarceration but the result of incarceration" (458).

The conditions in these institutions were caused by the inability or unwillingness of state governments to allocate sufficient funds for the care, education, and well-being of institutionalized children (Trent 1994, 256–257; Metzel 2004, 432). In the late 1960s, California Governor Ronald Reagan eliminated $17 million of funding for state hospitals and institutions for people with cognitive disabilities (Trent 1994, 256). In the late 1960s and early 1970s in New York, Governor Nelson Rockefeller spent $1.5 billion on a building project in Albany, while freezing hiring at the Willowbrook State School, which lost 912 of 3,383 of its employees by 1970 (Trent 1994, 258). Similar budget cuts occurred in other states, and during this period just prior to the lawsuits of the early 1970s, there was constant demand for new admissions to state institutions.

Litigation forced the states to release children housed in institutions, to create community housing and social programs for people with cognitive disabilities, and to increase funding for these initiatives (Rothman and Rothman 2004). The success of deinstitutionalization necessitated the creation of community supports for people with disabilities and their families where very little had existed before. Community support services for parents with children with Down syndrome, for instance, and arrangements for care in the community, often had to be created from nothing and expanded to meet the increased demand (Metzel 2004). Only a decade earlier many of these children would have been sent to institutions, so at the beginning of the deinstitutionalization process, many necessary community services simply didn't exist, or didn't exist on the scale needed. An example of this process is the inclusion of children with cognitive disabilities in public education. These children and their parents won the right to public education in the United States through constitutional challenges like *Mills v. Board of Education of the District of Columbia* (1972) and *Pennsylvania Association for Retarded Children v. Commonwealth of Pennsylvania* (1972). Federal legislation followed in 1975 with the Education for All Handicapped Children Act. The funding devoted to the inclusive education of children with disabilities increased eightfold in the years from 1976 to 1980—from $100 million to $800 million. The President's Committee on Mental Retardation described deinstitutionalization as a "total reorientation" in the care and education of people with cognitive disabilities (qtd. in Metzel 2004, 434). Housing, care, and education in the community are better alternatives to institutionalization—notably because large state institutions did not provide any of these necessities, but rather provided an environment marked by violence, disease, and neglect. Nonetheless, the change to a community orientation was a process that took time to establish services that could support parents with children who had cognitive disabilities.

In this context, the evidence provided by Kuhse and Singer in 1985, and the anecdotes they relate, might simply have reflected gaps in social services encountered by parents in the vanguard of the deinstitutionalization process. Many of the studies and works they cite date to the early days of deinstitutionalization in the 1970s. Kuhse and Singer appear entirely unaware of the massive changes in play during this period. By the late 1980s and into the 1990s and the first decade of the new millennium, many of the problems experienced by families with children having Down syndrome or other disabilities had been resolved or significantly mitigated, and the data on the functioning of such families became decidedly more positive as a result of improved social supports and greater acceptance of the community-based ethos of care. Seen from the perspective of this history, the recommendations that Kuhse and Singer make in favor of infant euthanasia seem out of keeping with the political and social changes in the care environment going on around them. The reasons that Kuhse and Singer give to support a policy of killing infants with Down syndrome sound similar to the reasons that parents were given in the 1950s to persuade them to institutionalize (and thereby brutalize) their children who had cognitive disabilities.

Conclusion

This chapter has outlined three possible responses to my use of positive assessments of family life in memoirs written by parents of children with Down syndrome and to my position that prospective parents should bring children with Down syndrome into the world and into their families. I have provided counterarguments to the first of these responses. According to this first response, it may be alleged that Down syndrome parenting memoirs are the product of self-deception or adaptive preference in which parents put a positive spin on their family lives as a way of coping with dysfunction. I have discussed some of the possible sources for this view. Some professionals and researchers believe that positive assessments of family life that involve a child with Down syndrome are inauthentic. Some bioethics literature—such as the book by Kuhse and Peter Singer (1985)—contains the view that a child with Down syndrome is almost always a source of dysfunction in a family.

To counter these views, I have argued that the allegation of adaptive preference directed at Down syndrome parenting memoirs is conceptually unfounded. Furthermore, I have presented extensive social science evidence that is either based on objective measures, or on validated self-report measures, showing that families that include a child with Down syndrome are no less functional than families with only typically developing children.

I have also given some historical context to Kuhse and Singer's claims about the destructiveness of bringing a child with Down syndrome into a family. Kuhse and Singer (1985) cite outdated literature from a time when deinstitutionalization was just beginning, and with improved research methods since then firmly in mind, there is little reason to believe that bringing a child with Down syndrome into a family today will have an adverse effect on family life.

5 The Wrongful Disability Objection

In August 2014, Oxford University scientist Richard Dawkins, claimed on twitter that it "would be immoral" to bring a child into the world who was diagnosed prenatally with Down syndrome (Kutner 2014). Dawkins had this recommendation for the prospective parents: "Abort it and try again" (qtd. in Kutner 2014). According to journalist Patricia E. Bauer (2005), whose daughter Margaret has Down syndrome: "Prenatal testing is making your right to abort a disabled child more like 'your duty' to abort a disabled child." This chapter addresses arguments that there is a moral obligation to abort fetuses diagnosed with Down syndrome.

In chapter 3, I presented memoirs written by parents of children with Down syndrome in order to argue that parents typically think that parenting such a child is positive and transformational. Chapter 4 addressed possible doubts about the veracity of such memoirs because of the phenomenon of "adaptive preference." To counter these doubts, I provided evidence in the form of social science research that the presence of a child with Down syndrome does not adversely affect family life, which shows that parenting memoirs are not distorted by adaptive preference.

The subject of this chapter is the argument that in spite of positive evaluations of raising a child with Down syndrome, it is nonetheless morally wrong to bring a child with a disability into the world when prospective parents could instead have a child without a disability. Dawkins is not the only advocate of this controversial position. In their influential book *From Chance to Choice* (2000), the philosophers Allen Buchanan, Dan W. Brock, Norman Daniels, and Daniel Wikler present an argument for the wrongness of choosing a child with a disability like Down syndrome.[1] According to this "wrongful disability" argument, it is morally wrong to give birth to a child with a disability when you could instead try to have a child without a disability, even when having the disability in question still allows a person to lead a worthwhile life. Buchanan and colleagues (2000, 243) identify Down syndrome as just such a disability—one that "will permit a worthwhile life." Philosophers Julian Savulescu and Guy Kahane (2009) also defend a variant of this position.

To address these arguments, I will outline the flaws in their reasoning and some of the mistaken assumptions they make about cognitive disability.

Wrongful Disability in *From Chance to Choice*

Buchanan, Brock and colleagues (2000) describe a thought experiment first devised by Derek Parfit (1984) in *Reasons and Persons*, often called the "non-identity problem." The thought experiment involves three cases, P1, P2, P3:

P1—"a woman is told by her physician that she should not attempt to become pregnant now because she has a condition that is highly likely to result in moderate mental retardation in her child. Her condition is easily and fully treatable by taking a quite safe medication for one month. If she takes the medication and delays becoming pregnant for a month, the risk to her child will be eliminated and there is every reason to expect that she will have a normal child. Because the delay would interfere with her vacation travel plans, however, she does not take the medication, gets pregnant now, and gives birth to a child who is moderately retarded" (Buchanan et al. 2000, 244).

P2—a similar condition is detected in a pregnant woman. However, if she takes the drug while pregnant, the child will not develop the cognitive disability. She does not take the drug, and the child is born with the disability (244).

P3—a child is born with a moderate cognitive disability, but it is treatable with a safe medication, which is likely to cure the child's condition. The child's mother fails to provide the medication to her child (244–245).

According to Buchanan and colleagues (2000, 244), most readers will have a strong commonsense intuition that the woman in P1 acts wrongly, "even if for pragmatic reasons many people would oppose government intrusion into her decision." They point out that P1 is very similar to P3, which many would understand as a case of medical neglect.

There is a problem, however, in trying to account for the wrongness of P1 using standard criteria of what makes something harmful. The child's disability in P1 is moderate and the child will be able to live a worthwhile life. The disability is not so severe that it is a fate worse than death. So the child in P1 is not harmed by being brought into existence. If the woman in

P1 were to take the medication and wait a month before conceiving, then a different child would be born. So unlike the medication in P2 and P3, the medication in P1 would not cure the prospective child's moderate disability. Instead the medication would allow for a different child to be born, one without a disability. Standard criteria of harm require that there is an individual who is harmed. But as P1 plays out, the woman does not harm anyone by bringing a child with a moderate disability into the world. Nonetheless, Buchanan and his colleagues (2000, 247) insist the commonsense intuition is that the woman in P1 has done something wrong "because of the easily preventable effect on her child." In order to solve this philosophical conundrum, they propose an innovation. Since traditional criteria of what makes something harmful cannot account for the intuition that P1 is wrong, Buchanan and colleagues propose a new criterion or principle of harm that includes the case of P1 as a case of harm. The traditional criteria of harm that depend on the existence of a harmed individual they describe as "person-affecting principles." To account for P1, they propose "N," a "*non*-person-affecting principle" of harm:

N: Individuals are morally required not to let any child or other dependent person for whose welfare they are responsible experience serious suffering or limited opportunity or serious loss of happiness or good, if they can act so that, without affecting the number of persons who will exist and without imposing substantial burdens or costs of loss of benefits on themselves or others, no child or other dependent person for whose welfare they are responsible will experience serious suffering or limited opportunity or serious loss of happiness or good (Buchanan et al. 2000, 249).

Principle N makes what constitutes harm dependent upon a comparison between the prospective suffering, happiness, opportunity, or good of two children. Harm exists if child 1 is born when child 2 could have been born instead with better prospects for these benefits. The principle is simply a restatement, in formal terms, of the intuition supposedly generated by P1. It does not explain how it is possible to cause harm without any victims. The authors assert nonetheless that this form of harm is possible.[2]

In another article, Buchanan's coauthor Dan Brock writes, "non-person-affecting principles in effect reject the need for an answer to the question, for *whose* sake should a disability in cases like P1 be prevented" (Brock 2005, 87). Because it harms no one to bring a child into existence who will have a life worth living, Brock resorts to an impersonal perspective. He says, "It is for the sake of a world with less diminishment of well-being or limitation of opportunity" that the child with the disability in P1 should not exist (Brock 2005, 87). Buchanan, Brock, Daniels, and Wikler (Buchanan et al. 2000) do not advance the (offensive) argument that we would all be better off if children with moderate disabilities did not exist. They only mean to describe a harm in principle—the abstract idea that there exists less opportunity and well-being in the world when such children are not replaced by others with putatively better prospects.

Even though P1 is an odd scenario unlike most real situations in which prospective parents must decide whether to bring a child with a disability like Down syndrome into the world, Buchanan and colleagues clearly believe that this thought experiment can be applied to such situations. The authors deliberately phrase principle N in such a general way that it applies to cases of prenatal testing and selective abortion. They discuss exceptions to the judgment that it is wrong to give birth to a child with a disability in real-life cases. Principle N accounts for such exceptions with a clause that excuses parents who bring a child with diminished opportunity into the world when attempting to have a different child would impose "substantial burdens or costs or loss of benefits on themselves or others" (Buchanan et al. 2000, 249). Among such "substantial burdens" are "possible moral objections by the parents to the use of abortion" (250). The authors also believe that P1 and the N principle to which it gives rise have implications for real-life genetic counseling practices. The authors state that nondirective genetic counseling—in which the counselor stays neutral about the question of whether the pregnant woman should continue the pregnancy or terminate—is "morally problematic" (256). Buchanan and colleagues apparently believe that genetic counselors have an obligation to advise some pregnant women to terminate their pregnancies because doing otherwise will diminish the amount of opportunity and well-being in the world.

Rejecting the Intuition

One obvious response to P1, and the argument that follows from it, is to reject the "commonsense" view that the woman acts wrongly by choosing to have a child with a moderate disability. The underlying rational procedure being used in the construction of principle N following from P1 is "wide reflective equilibrium" (Buchanan et al. 2000, 22). As the authors describe it, wide reflective equilibrium is a form of systematic moral reasoning in which one attempts to arrive at consistency between (a) the moral judgments we have about concrete and hypothetical cases, (b) the moral principles to which we subscribe, and (c) underlying theories about "how the world is and how people in it behave" (Buchanan at al. 2000, 22). In order to arrive at consistency among judgments about cases, moral principles, and underlying theories, we must sometimes revise or dispense with any one of these elements if there is an inconsistency in our beliefs. We can see the authors use this procedure when they revise the person-affecting principle of harm to produce the non-person-affecting principle N, which better accommodates P1. The "commonsense" intuition that the woman in P1 behaves wrongly is not consistent with a person-affecting principle of harm, so to achieve consistency, they revise the principle into one that is consistent with non-person-affecting cases.

However, rather than treating the moral judgment generated by P1 as authoritative, and revising the person-affecting principle in order to arrive at equilibrium, rejecting the supposed "commonsense" judgment in P1 is another option. Buchanan and colleagues provide little or no reason for accepting their revised principle (principle N) apart from the P1 "commonsense" intuition. A rejection of the P1 intuition would create equilibrium without needing to revise the person-affecting principle of harm. A virtue of this strategy is that one can preserve the compelling belief that causing harm requires that someone be harmed.

According to the theory of wide reflective equilibrium, intuitions (moral judgments about cases) are just as revisable as moral principles (Daniels 1996; McMahan 2013; Brink 2014). There are two levels at which intuitions are said to be revisable. First of all, in order to enter into the balancing process with moral principles and background theories, intuitions must qualify as "considered moral judgments." Immediate, pretheoretical, "gut-feeling" reactions, in and of themselves, do not qualify as considered moral judgments (Daniels 1996; Brink 2014). To qualify as considered moral judgments, intuitions must pass through a kind of filter meant to weed out judgments that are the product of misinformation, incapacity, or obvious forms of bias. For example, an intuition formed in haste, without reflection, may be rejected after a more considered appraisal.

The second level of revisability is the process of balancing that I have described. It occurs when considered moral judgments are placed alongside relevant moral principles, and relevant background theories of society and human behavior. When there is inconsistency among these three elements (judgments, principles, theories), considered moral judgments can be rejected in order to produce equilibrium, as can principles or theories. Considered moral judgments have only defeasible epistemic authority (Daniels 1996; McMahan 2013; Brink 2014).

There are several reasons for rejecting the P1 intuition that the woman behaved wrongly. One important reason has to do with the influence of ableist bias over the P1 thought experiment. People who are not disabled might be inclined to believe that living with a disability invariably causes diminished well-being, and thus that it is wrong to bring a child with a disability into the world when this can be avoided. The bias of ableism could be viewed as a distorting effect that causes the "commonsense" intuition in P1 to be "filtered out" in the first instance so that the intuition fails to qualify as a "considered moral judgment." Alternatively, if the P1 intuition does qualify as a "considered moral judgment," a realistic background theory about the nature of bias and the ability of bias to influence one's beliefs about ethics could inform the balancing process of wide reflective equilibrium so that the authority of the P1 intuition is recognized to be suspect. Such an understanding of bias would seem to be one of the crucial background theories about "how the world is and how people in it behave" (Buchanan at al. 2000, 22) useful for wide reflective equilibrium.

Some findings of empirical research in moral psychology also give us reason to reject the P1 intuition. For instance, "framing effects," which may be present in the language used to describe cases, in the emphasis or downplaying of certain elements of a case, or in the order in which options appear in a case, can affect the intuitions that readers form about a case (Brink 2014). In *From Chance to Choice*, Buchanan and colleagues seem to deliberately employ framing effects to generate the intuitions that they want, rather than trying to avoid misleading their audience with these techniques. Thus, in presenting case P1 alongside cases P2 and P3, they deliberately frame P1 in such a way that readers arrive at the intuition that the woman in P1 behaved wrongly. Recall that P2 is a case in which a woman refuses a treatment while pregnant that would cure a fetus's disability, and P3 involves denying a treatment to an infant. They claim that the nonpregnant woman in P1 behaves "no different morally" than in P2, and they state "[n]or is [P1] different morally" than P3 (Buchanan et al. 2000, 244). Equating P1 with P3, the authors state that "her action in P3 would probably constitute medical neglect" (245). Even though there are differences between the cases (such as those between not being pregnant, being pregnant, and being responsible for an infant) that readers might consider to be morally relevant, Buchanan and colleagues effectively frame P1 as an example of child abuse.

Furthermore, they rhetorically construct P1 to make the choice seem simple, the outcomes certain, and the reasons for refusing to wait a month are made to look trivial in comparison—"the delay would interfere with her vacation travel plans" (Buchanan et al. 2000, 244). Such framing tactics prevent the P1 intuition from qualifying as a considered moral judgment. The authors use the P1 thought experiment as a rhetorical tool to support a position in an argument about the wrongness of bringing a child with a disability into the world, and once we recognize these rhetorical techniques, we ought to question the intuition that the woman in P1 behaves wrongly.

It is also worth pointing out that moral intuitions even about supposedly noncontroversial cases can vary (Brink 2014). For instance, Carol Gilligan (1982) has famously showed that gender-based differences in moral reasoning can affect intuitive judgments. Buchanan and colleagues (2000, 244) refer to the P1 intuition as arising out of "commonsense moral views," and "commonsense morality." But this "commonsense" intuition is not commonsensical to everybody. Elizabeth Barnes (2016) and David Wasserman (2005), among others, have expressed disagreement with the supposed intuition that the woman in P1 behaves wrongly. It is fairly easy to believe that since the child with the moderate disability has a worthwhile life, no harm is done by becoming pregnant and giving birth. This variability further suggests that the P1 intuition is the product of bias.

Another problem is the exoticism of P1. The scenario described in P1 is unlike realistic cases of reproduction. P1 does not resemble any currently realistic clinical scenario. But Buchanan and colleagues have nonetheless used the thought experiment to generate an "authoritative" intuition that they claim can be applied to more common and realistic reproductive situations. The problem is that intuitive reasoning in various domains works best when we are dealing with familiar features of the world or with common experiences (Kahneman 2011; Brink 2014). We are able to make accurate judgments at a glance when our judgment has been conditioned by repetition and familiarity. On the other hand, when we deal with something exotic or novel, it makes more sense to take our time, engage in analysis, and access slower processes of thought. As the saying goes: "Hard cases make bad law." Since P1 is an exotic case, we should not treat it as authoritative when addressing more familiar reproductive situations. When one compares P1 to real-life reproductive situations, few would conclude that it is wrong for prospective parents to bring a child with a moderate disability into the world when they could selectively terminate and try again for a nondisabled child. I will return to this line of argument later on in the chapter. The "authoritative" intuition Buchanan and colleagues think is generated by P1 becomes even less certain when we alter P1 so that it resembles a more realistic scenario.

Opportunity and Well-Being in *From Chance to Choice*

Recall that Buchanan and colleagues' wrongful disability argument is predicated on the view that children born with a cognitive disability like Down syndrome will have diminished opportunity and well-being compared to children without disabilities (Buchanan et al. 2000, 278). For the most part, these authors take this comparison to be obvious and do not offer much in the way of support for it. However, an examination of their discussion of both concepts—well-being (or welfare) and opportunity—gives reason to be skeptical of the claim that lives of people with cognitive disabilities have these deficiencies.

Well-Being

Empirical evidence into the lives of people with various disabilities shows that people with disabilities tend to rate their quality of life higher than nondisabled people would rate a life with a disability (National Organization on Disability 1994; Sackett and Torrance 1978). In fact, self-assessed quality of life by people with many disabilities—even severe disabilities—are usually comparable to self-assessments made by people without disabilities (Schroeder 2016; Ubel et al. 2005; Albrecht and Devlieger 1999). For example, in a recent survey I have discussed in earlier chapters involving 284 people with Down syndrome, 99 percent reported that they were happy with their lives, and 97 percent indicated that they liked who they are (Skotko et al. 2011). Among the authors of *From Chance to Choice*, Dan W. Brock (2005, 73; Brock 1993) acknowledges an awareness of these data and similar findings. Brock argues that part of what explains the high self-assessed quality of life of people with disabilities is the phenomenon of adaptation—and related concepts of coping and accommodation—in which a person sets or adjusts life plans to reflect the reality of their disability, or develops skills in other areas of their lives to compensate for disabling features (Brock 2005, 73).

Even though social science research has clearly established that self-assessments by people with disabilities are, in fact, valid reports of how well they believe their lives are going, in trying to support the view that people *without* disabilities tend to have better well-being, Brock contends that subjective assessments of quality of life by people with disabilities do not reflect their true quality of life. First of all, Brock (2005, 74) argues that some people with disabilities do not adapt. For such people, "both they and others would rate their quality of life as significantly diminished by their disability" (Brock 2005, 74). Brock notes as well that the process of adapting to a disability can be burdensome (Brock 2005, 74–75). The problem with this line of argument is that the data on self-assessment of quality of life by people with disabilities contradict these claims. To be sure, some people with disabilities may not adapt and may have a diminished quality of life. But this fact places them in the company of all other people with any kind of challenge or disadvantage in life. Poverty, exposure to violence, illness unrelated to disability, unhappy family relationships, and loneliness can all be factors that would threaten a person's quality of life. Some people adapt to these challenges, and some do not. In many cases, overcoming factors like these can be burdensome. The point of the data on self-assessment of quality of life is that *on average* people with disabilities do not find it more difficult to overcome the challenges in their lives than nondisabled people do.

Brock's second line of argument is that a person's subjective assessment of quality of life by itself is not a complete indicator of that person's true quality of life. According to Brock: "The *opportunity for choice* from among a reasonable array of life plans is an important and independent component of quality of life: it is insufficient to measure only the quality of the life plan the disabled person now pursues and his or her satisfaction with it" (Brock 1993, 124; emphasis original). The claim here is that even if you have a disability like Down syndrome and give a positive self-assessment of your quality of life, this is not a good indicator of your quality of life because your disability limits opportunity in the choice of your life plan. Brock's reasoning here could be considered a variant of an "open future" argument—that parents have an obligation to provide their children with a future that has as many opportunities as possible.

Brock offers very little in support of this view, other than to point out that opportunity is a factor that might contribute to a person's quality of life. To explain, he gives the examples of congenital blindness and deafness. People born blind or deaf may rate their quality of life high and would not even know what they are missing by not being able to see or hear. But in the case of a person's blindness, "there will be valuable human activities requiring sight that will not be possible" (Brock 2005, 75)—hence their quality of life is lacking in comparison to someone who can see. For someone who is deaf, "there are still valuable human activities, such as the appreciation of music, that are closed off to deaf persons" (Brock 2005, 75). According to Brock, this inability to engage in valuable human hearing activities diminishes a deaf person's quality of life, even if the deaf person rates his or her quality of life as high as a hearing person. Brock mentions the deaf community critique that, in the context of a rich and vibrant deaf culture, being unable to hear is not a deficiency in quality of life. He shrugs off this critique and asserts that the "richness" of life is more available to those who can see or hear than to those who cannot (Brock 2005, 75). Brock's response to this critique by the deaf community illustrates that his argument amounts to little more than a brute assertion that the opportunity to choose among a wide array of life plans must be a factor in the quality of a person's life.

As we have seen, the wrongful disability argument arising out of the P1 thought experiment depends on the idea that people with cognitive (and other) disabilities like Down syndrome have less opportunity and diminished well-being than people without disabilities. Buchanan, Brock and colleagues claim that people with disabilities have diminished well-being to a large extent because such people have diminished *opportunity*. The concept of opportunity emerges as the key to the wrongful disability argument. Their claim about well-being can be reduced down to their claim about opportunity. Without the assertion that people with disabilities have diminished opportunity, there is very little reason to believe that they have diminished well-being as compared to people without disabilities.

Opportunity

In *From Chance to Choice*, Buchanan and colleagues present a particular understanding of the concept of opportunity. They understand opportunities as phenomena that are constituted by one's social environment. The authors call this social environment "the dominant cooperative framework of a given society" (Buchanan et al. 2000, 79). This Rawlsian concept is defined as "the set of basic institutions and practices that enable individuals and groups in a given society to engage in ongoing mutually beneficial cooperation" (Buchanan et al. 2000, 79). Buchanan and colleagues have in mind that this framework takes a specific form in contemporary society: "In the United States and other 'developed' societies, the most basic cooperative framework consists, to a large extent, of the competitive market system" (Buchanan et al. 2000, 259). It is within this economic context that Buchanan and colleagues judge the woman's choice in P1 and in which they understand the concept of "opportunity" in their principle N. They consider the child with the moderate disability who is born in P1 to have less opportunity (and thus lower well-being) because he or she will not function as well in a competitive capitalist market system. A defining feature of opportunity for Buchanan and colleagues is "successful participation in the industrial economy" (Buchanan et al. 2000, 289). In some places, the authors write about opportunity in the context of choices among a broad array of life plans. But often their discussion of opportunity highlights a narrower range of human interactions—those concerned with economic competitiveness.

The paradigm example of someone with an optimal level of opportunity in this "dominant cooperative framework" is the white-collar worker. Buchanan and colleagues argue that they devalue disabilities themselves, without devaluing the people who have disabilities. They illustrate this position with an anecdote: a parent encourages a child to do well in school in order to increase her opportunity and avoid the need to settle for "menial" or "blue-collar" jobs (Buchanan et al. 2000, 278–279). The parent understands that doing well in school is a prerequisite for getting a "white-collar" job. The authors acknowledge that we should not ignore the opportunities for skills development and job satisfaction that can be found in blue-collar employment. But they go on to state that "Nevertheless, the advice the parent gives may be sound … in the light of a realistic estimate of the educational and economic facts of life, without in any way denigrating persons who perform menial labor" (Buchanan et al. 2000, 278–279). Just as one can devalue blue-collar labor without devaluing blue-collar laborers, one can devalue disabilities without undermining the value of people with disabilities. If the point of their anecdote is transferred to the reproductive context, the educational and economic facts of life dictate that a pregnant woman should selectively abort a fetus with a cognitive disability and try instead for a child without a disability, who is more able to become a white-collar worker, thereby maximizing the child's opportunity. But Buchanan and his colleagues believe that this position does not denigrate anyone who has such a disability.

It is a fairly disappointing result if the wrongful disability argument in *From Chance to Choice* comes down to this: pregnant women informed their prospective children will have a disability should terminate their pregnancies because people with these disabilities typically cannot do white-collar jobs. If this is the reasoning that stands behind the supposedly "authoritative" intuition that P1 is a case of wrongful disability, then it seems that such an intuition is really just a reflection of class-based bias and of a desire to ensure that children are able to fit well into our competitive economic system. But human life is not so simple. Economic competitiveness is only one value among many. It is not the whole of society. And the societies that prioritize this one value over all others may not be very humanitarian. Aside from participation in the competitive market, a child with a disability like Down syndrome will be born into a family, into a community, into cultures, into a nation-state that is defined by laws rather than exclusively by market forces. He or she will become involved in social practices and relationships that are not defined by the market—relations of love, friendship, citizenship, and many others. Even if people with Down syndrome do not typically become optimal participants in the market economy (though many do), they and others with disabilities can make small and large contributions to social life within these other meaningful areas of human association.

Here is an example: the 2013 documentary *The Crash Reel* tells the story of snowboarder Kevin Pearce's traumatic head injury suffered while training for the 2010 Winter Olympics, and his recovery from the injury. Pearce eventually adopts the goal of returning to the sport that he loves, and that almost killed him. The documentary is also the story of Pearce's family, who supports him throughout his recovery, but opposes his return to snowboarding because of the risk of future injury. Even a small blow to his head could be devastating. Kevin Pearce's brother David has Down syndrome. Throughout the documentary, David shows a moving degree of insight, sympathy, and love toward his brother. David is an integral member of the family who involves himself in the emotional work of helping his brother recover, and of advocating for the position that Kevin should not return to the slopes. David engages in the difficult and sensitive task of trying to balance attention to his brother's psychological well-being against a sense that Kevin's strongest wish should be overruled. This task cannot be accomplished without large reserves of trust, and confidence in one another, and among the Pearce family members. David Pearce shows that people with Down syndrome are clearly able to make meaningful contributions to social life. The wrongful disability argument in *From Chance to Choice* fails to count these social contributions as valuable because of its undue emphasis on opportunity, understood as economic competitiveness.

Those who support Buchanan and colleagues' wrongful disability position would counter that such social contributions are certainly not unique to people with Down syndrome. Typically developing family members can also demonstrate psychological insight and sympathy. That being the case, they would argue that nothing is lost through selective abortion and trying again for a nondisabled child who could make the same social contributions as a child with Down syndrome. But this objection concedes an essential point: nothing is lost, but nothing is gained either, except a dubious calculation of economic competitiveness. People with Down syndrome are just as able to contribute to important aspects of social life as people without disabilities. On this basis, there can be no moral obligation to selectively terminate a pregnancy and try again for a child without a disability when given a prenatal diagnosis of Down syndrome.

Comparative Judgments

It is a fact that children and adults with Down syndrome tend to have some degree of cognitive disability and a consequent increased level and duration of dependency. People with Down syndrome typically require assistance with activities of daily living for a longer period of time during the span of their lives than people who do not have disabilities. There is a strong likelihood that a child born with Down syndrome will have diminished economic opportunities as compared to those who are born without this or a similar condition. Another likely consequence of cognitive disability and increased dependency is difficulty in achieving common milestones of a successful middle-class life—like getting married or attending university. As we have seen, this concern is also cited by authors of parenting memoirs.

Though Buchanan and colleagues place undue emphasis on economic opportunity, prospective parents might think that challenges in realizing life milestones and lower economic opportunities would give a child with Down syndrome lower well-being as compared to a child who does not have these challenges, even though these challenges do not reduce well-being in an absolute sense. I have pointed out limitations in the way that Buchanan and his colleagues present the concepts of well-being and opportunity. We should also examine whether there can be an obligation to selectively terminate on the basis of comparative judgments of well-being that take into account the consequences of dependency. We want our children to thrive and do well in their lives. Going to university, getting married, and making money are commonplace signs that a person is doing well.

As noted in chapter 3, many people with Down syndrome do get married, and some go to university, though they are less likely to do so than people who do not have this condition. Cognitive disability and increased dependency are not differences that unilaterally result in an inability to realize these life milestones. Among the life milestones I would also include smaller and larger achievements, such as report cards with straight A's, significant successes in sports, or in music, or other life pursuits. People with Down syndrome can realize goals in all of these areas, sometimes within disability-specific organizations such as the Special Olympics. When it comes to the effects of disabilities on life milestones, it is obvious that there is no single life milestone, and no set of milestones together, that are necessary for one's life to go well. There are happy and thriving people who are single and who have not graduated from university. Achieving a milestone such as home ownership is not a requirement for happiness. Well-being is not contingent on having children or grandchildren. We can be happy without being straight-A students or marathon winners. We can live good lives without achieving any of these or any other milestones.

Although extreme poverty can compromise one's well-being, above a certain income threshold, having more money does not translate into greater happiness or greater emotional well-being (Kahneman and Deaton 2010). We don't have to be rich or even members of the upper middle class for our lives to go well. Furthermore, those of us with a disability can avoid poverty (and the risk to well-being posed by poverty) in a number of ways. Personal income does not always have to come from paid employment in order to avoid the negative effects of poverty. Social programs or family savings can keep people with a disability out of poverty. In other words, being a success in our capitalist market system, or even finding employment in it, is not a necessary condition for well-being.

These observations about milestones and economic opportunity show that achievements of various kinds and economic success are not determinants of well-being. That said, a comparative judgment of well-being might run as follows:

1. Even though a child who has Down syndrome will not have a bad life because of diminished prospects for achieving certain life milestones and economic opportunity, a child who does not have Down syndrome will have better prospects for achieving these things.
2. Even though a child with Down syndrome will likely not have a low level of well-being, a child without Down syndrome will have a chance to have an even higher level of well-being due to the greater likelihood of achieving life milestones and economic opportunity.
3. Prenatal testing and selective abortion could enable one to "try again" for a child without Down syndrome.
4. Because the nondisabled child will have a better chance for higher well-being, one must selectively terminate and "try again" if given a prenatal diagnosis of Down syndrome.

To accept that they have such an ethical obligation, prospective parents would have to also accept at least two key assumptions. First, they would have to assume that the purpose of parenting is to maximize the well-being of their children, even if this means terminating a previously wanted pregnancy and attempting to have a different child. A second assumption is that well-being can be maximized through achieving life milestones and through enhanced economic opportunities. These assumptions, in turn, constitute a particular view of parenthood: that the role of the parent is to facilitate, through virtually any means, the maximization of a child's achievement in the social and economic spheres. I will call this the "parent as achievement maximizer" view. An important feature of this view of parenthood is that parents must take extreme measures—such as ending a wanted pregnancy in an attempt to have a different child—in order to maximize the achievements of their children.

But the goal of maximizing achievement in our children is either inconsistent with a necessary condition of good parenting, or at least runs the risk of being inconsistent with this necessary condition. The necessary condition is Kantian. In any defensible conception of good parenting, we must value a child for the child's own sake. This Kantian principle underlies human rights more generally: persons ought to be valued as ends in themselves rather than only as means to an end. One cannot value a child for her own sake prior to her existence, but when she is brought into existence, a child just like any other person should be recognized to have intrinsic value. Christine Overall (2012, 88) formulates the Kantian necessary condition in this way: "In procreating, the parents should undertake to create a being whom they will value for his own sake."

Thus a parent who values a child in strictly instrumental terms is a kind of monster. The child who is seen only as a laborer, or only as "another mouth to feed," or who is a victim of sexual abuse, or the child who is not mourned or missed in death, is not valued for her own sake.

The maximization of a child's achievement can also lead to, or be motivated by, a failure to accord the child the intrinsic value he or she deserves. Consider children who are driven to succeed in sport, or in music, or in other areas of life, through intensive training and constant pressure. In some cases, a parent is motivated to drive a child to succeed because the parent failed to be successful in the same pursuit in his or her own life. In such a case, their attempts to "improve" their children can be so harsh and unremitting and so bereft of affection that it is clear that they are using their children as means to their own ends—as instruments to achieve the success they themselves could not, for example. Though this example is extreme, it shows that the parent as achievement maximizer view can be inconsistent with the Kantian necessary condition of good parenting. It is possible to pursue achievement (in life milestones, in economic success, or in other ways) in a way that values the child only as a means to an end. This example alone shows that the parent as achievement maximizer view is flawed. There is something wrong with the idea that a parent's role is to maximize a child's achievements in social and economic terms.

In addition to achievement maximizing parents who clearly violate the Kantian necessary condition, there are also cases that run the risk of violating this condition, though they might not be clearly in violation. The problem is that fulfillment of the Kantian necessary condition requires a degree of acceptance of a child for who she is, regardless of her achievements, and there is tension between the pursuit of achievement and acceptance of the child for who she is. To illustrate, Janis Gonzales, one of the memoirists cited in chapter 3, writes, "Perhaps I was genetically burdened with high self-expectations, or possibly my parents and teachers accidentally gave me the impression that my entire value as a person came from getting A's, all the time" (Gonzales 2007, 249). As a child, Gonzales felt that she was valued conditionally—not for her own sake, but for what she could accomplish (straight A's). She attributes her insecure perfectionism throughout medical school and in professional life as a doctor to her childhood experiences. But Gonzales's values changed when her daughter Cariana was born and was diagnosed with Down syndrome.

Gonzales's parents and teachers may actually have seen her as valuable for her own sake, regardless of what Gonzales herself believed. The impression they left about Gonzales's value as a person may have been truly accidental. But, even so, parents' pursuit of maximum achievement always runs the risk of violating the Kantian necessary condition of good parenting because it matters whether the child herself feels that she is valued unconditionally. It is enough of a violation that Gonzales is left with the impression that her value as a person was contingent upon academic achievement. It is difficult to argue that a child is valued for her own sake if the child herself does not feel valued for her own sake. Furthermore, a child's well-being may be affected if she has doubts about whether her parents love and accept her—if she believes she must perform in order to receive that acceptance. So in addition to the parents who clearly violate the Kantian necessary condition of good parenting, there are many who, in pushing their children to achieve life milestones, run the risk of violating the Kantian condition, even though they might believe they are valuing their children for their own sake.

For these reasons, the parent as achievement maximizer view is highly problematic. Of course, many parents are legitimately concerned about the success of their children, and successes can contribute to well-being. As such, it would be wrong to criticize much of what parents do to help their children realize achievements in their lives. Many parents recognize when they are pushing their children too hard and back off when it is clear this is negatively affecting the children. My criticism of achievement maximizers is directed not at such parents but only at the view that parents have a duty to *maximize* achievement—by going to great lengths and taking extreme measures— above all else, often at the cost of fulfilling other parental duties. I would categorize the choice to selectively terminate in order to "try again" for a child without a disability as falling under this critique. The termination of a wanted pregnancy on the basis of a judgment comparing the prospective achievements of two potential children would qualify as an extreme measure. The comparative judgment resulting in selective termination is an extreme measure because it results in preventing the birth of a child whose life would likely go well simply to have a child with a greater ability to achieve life milestones and economic opportunities.

As my analysis reveals, the judgment comparing the well-being of a child with Down syndrome to the well-being of a nondisabled child on the basis of the ability to achieve life milestones and economic opportunity provides little reliable insight into parental duties or the well-being of children. In contrast, parents of children with Down syndrome, like Janis Gonzales, often describe a powerful norm of acceptance as the cornerstone of their relationship with their children (Hampton 2012, 54; Foraker 2009, 38; Groneberg 2007, 292; King et al. 2006, 361). This norm of acceptance is a full expression of the Kantian necessary condition of good parenting. Gonzales sees her daughter Cariana as valuable for her own sake.

Savulescu and Kahane: The Principle of Procreative Beneficence

Thus far, my argument against the objection that it is morally wrong to bring a child diagnosed with Down syndrome into the world has focused on the version of this position defended in *From Chance to Choice* and elsewhere by Buchanan, Brock, and colleagues. In another and similar version of this position, Julian Savulescu and Guy Kahane argue that prospective parents have a moral obligation ("a significant moral reason") to select a child who will enjoy "the most well-being." This supposed obligation supports what they call the "Principle of Procreative Beneficence," which applies to procreative choices involving the child's genetic endowment, such as choosing between embryos to implant through the use of preimplantation diagnosis after in vitro fertilization and prenatal testing with the option of selective abortion (Savulescu and Kahane 2009; Savulescu 2001).

Numerous authors have argued that prospective parents are not ethically bound to abide by such a principle (e.g., Bennett 2009; de Melo-Martín 2004; Elster 2011; Herissone-Kelly 2006; M. Parker 2007; Sparrow 2007, 2011; Stoller 2008). Savulescu and Kahane (2009) address a number of these critical positions. For the most part, my objections to the Principle of Procreative Beneficence differ from those offered by other authors. My views are most similar to those who argue that there are insurmountable problems associated with the idea that a list of genetic traits can give prospective parents any reliable information about how well a prospective child's life will go (see especially de Melo-Martín 2004; M. Parker 2007; Sparrow 2007).

Savulescu and Kahane (2009, 284–289) generally argue that the Principle of Procreative Beneficence requires selecting against disability. But they propose an idiosyncratic definition of "disability." According to their "welfarist" definition, disability is

> a stable physical or psychological property of subject S that (1) leads to a significant reduction in S's level of well-being in circumstances C, when contrasted with realistic alternatives, (2) where that is achieved *by* making it impossible or hard for S to exercise some ability or capacity, and (3) where the effect on well-being in question *excludes* the effect due to prejudice against S by members of S's society. (Savulescu and Kahane 2009, 286)

On this view, they define "disability" as any lack of capacity or ability that lowers well-being (286). If Savulescu and Kahane were to accept my arguments and the empirical research findings about the well-being of people with Down syndrome, then, according to their definition, Down syndrome would not qualify as a disability. Down syndrome does not lead to a reduction of well-being in the way required by the definition for it to qualify as a disability. Nonetheless, comments in other parts of the article indicate that Savulescu and Kahane (2009, 283) believe Down syndrome causes people with this condition to be less "advantaged" and thus to have lower well-being than people without the condition or other disabilities. Furthermore, the choice to define disability in terms of a reduction of well-being suggests that these authors believe that physical and cognitive disabilities tend to lower well-being. It would be odd to define disabilities in this way if they did not think that their definition captures most characteristics commonsensically thought to be disabilities.

Savulescu and Kahane's overall argument assumes both that a person's well-being is a function of that person's intellectual and affective capacities and that these capacities are encoded in the person's genes. The purpose of choosing the potential child with higher capacities is that such a child can be expected to have "the most well-being" (Savulescu and Kahane 2009, 274).

I find Savulescu and Kahane's assumptions and therefore also their position to be simplistic in several important respects. First of all, a high level of functioning in capacities that many think would be beneficial does not always translate into high levels of well-being. Take memory, for example. Savulescu and Kahane consider "the capacity to remember things better" as a trait that clearly contributes to a better chance at a good life (Savulescu and Kahane 2009, 284). Buchanan and colleagues (2000, 168) would seem to agree, finding "a very substantial increase in the capacity for memory" to be a "general-purpose benefit." But, as we shall see in the following section, this is not always the case.

Memory and Well-Being

Our memories enable us to see ourselves as continuous beings maintaining a single identity over the passage of time. Indeed, the dominant theory of personal identity since the time of John Locke in the seventeenth century holds that continuity of identity is a function of psychological linkages, such as memories, between the past, present, and future (Locke [1690] 1979; see also Parfit 1984). Without an adequate ability to remember, we would lose our sense of self and become isolated in time from the life that we have lived. A good memory is clearly valuable. Because of the way our society rewards intellectual talent, having a better memory might seem an obvious benefit—especially for those engaged in intellectual labor of one kind or another. The quick-witted individual who can easily remember useful facts is generally regarded as more intelligent than others who cannot.

But there is such a thing as a harmful excess of memory, since it may correspond to an insufficiency in the capacity to forget. There are two related problems that could be caused by memory enhancement, each arising from the possibility that a greater ability to remember may weaken the ability to forget. The first problem is the emotional impact of being unable to forget events that are better left unremembered. Friedrich Nietzsche stressed the importance of forgetting, which he found to be "no mere *vis inertiae* [passive faculty] as the superficial imagine; it is rather an active and in the strictest sense positive faculty.... The purpose of active forgetfulness ... is like a doorkeeper, a preserver of psychic order, repose, and etiquette: so that it will be immediately obvious how there could be no happiness, no cheerfulness, no hope, no pride, no *present*, without forgetfulness" (Nietzsche [1887] 1967, 57–58; emphasis original). A capacity for memory that inhibits forgetting can be a plague to those who have experienced trauma. Posttraumatic stress disorder (PTSD) affects large numbers of people—from soldiers who have lived through combat, to people who have survived car accidents or violent crimes. Recent research into the effects of drugs such as propranolol has shown that it is possible to blunt the psychological effects of trauma through pharmaceutical interventions that affect memory systems in the brain (Glannon 2006). These interventions promise to remove or minimize the negative psychological effects associated with traumatic memories.

The idea that drugs could be used in the treatment of memories that cause PTSD has been met with bioethical debate about whether such research should be pursued, and whether such treatments, if realized, should be made available to those who have experienced trauma (Glannon 2006; President's Council on Bioethics 2003; Henry, Fishman, and Youngner 2007). The debate has introduced terms such as "therapeutic forgetting" and "pharmacological memory dampening" into the academic vocabulary (Cadwallader 2013; Chandler et al. 2013). These terms are suggestive of the harms associated with maximizing the ability to remember. It can be therapeutic to forget. It may be beneficial to dampen intrusive and unwanted memories. If enhanced memory impairs our ability to engage in a healthy level of forgetfulness, our emotional well-being may suffer.

The second problem associated with enhancing memory is that excessive memory can impede the efficient functioning of the brain by cluttering our minds with trivial and useless facts. William James recognized the importance of forgetting for good psychological functioning, stating that, "Selection is the very keel on which our mental ship is built. If we remembered everything, we should, on most occasions, be as ill off as if we remembered nothing" (James 1890, 680). The image of a keel is instructive, since it implies balance. A greater ability to remember may upset the delicate balance between the complementary faculties of remembering and forgetting. Just as the excessive forgetfulness of dementia can threaten our personal identity, so excessive memory can threaten our mental equilibrium. Diagnosed with "hyperthymestic syndrome," Jill Price has perfect recall of every day of her life from the age of eleven on (McGaugh 2013). But in her book *The Woman Who Can't Forget*, Price (2009) tells us that, far from improving her well-being, her perfect recall has made it hard for her to think clearly, to live in the present and plan for the future, and has led to physical and mental exhaustion.

Mistaken Assumptions about Well-Being

But memory is just one example given by Savulescu and Kahane of an ability whose maximization might contribute to the increased well-being of children. That they are wrong about the certain benefit of enhanced memory doesn't mean they are wrong in their overall defense of the Principle of Procreative Beneficence. But there is a more fundamental flaw in their arguments for this principle, and the problems associated with enhancing memory reflect this flaw. Savulescu and Kahane seem to assume that if prospective parents can simply max out a child's cognitive and psychological potentials, then the child will have a better chance at a good life. The problem is with the idea of maximization. Take the capacity for concentration as an example. Savulescu and Kahane suggest that concentration, along with memory, empathy, self-control, sociability, and intelligence are among the capacities whose maximization could be conducive to increasing well-being (Savulescu and Kahane 2009, 280, 282). The problem with maximizing concentration is similar to the problem with enhancing memory. In an article about the ethics of psychopharmacological enhancements, Walter Glannon cites the case of a woman who was administered drugs that enhance concentration as a treatment for concussions she had experienced. When asked what it was like to be treated with concentration-enhancing drugs, the woman stated that, "I worked like a demon, but I found myself disconnected. At the computer I was entirely focused, but off duty, certain pleasures, like wandering around aimlessly in my own mind, were no longer available to me" (Glannon 2006, 77). The increased ability to concentrate had a side effect of diminishing the woman's other cognitive capacities, such as the ability to let her mind wander.

My general point here is that just as there is no risk-free drug, any attempt to maximize a cognitive or psychological capacity will entail a trade-off that may not be beneficial for well-being. A high level of functioning in any of the capacities that Savulescu and Kahane identify as obvious contributors to high degrees of well-being could have an unanticipated downside. As Aristotle pointed out, sometimes an excess can be just as problematic as a lack. An excess of empathy could cause us to be so other-regarding that we suffer through self-neglect or through too great a sensitivity to the evils of the world. An excessive ability to concentrate may cause us to lose the substantial pleasures of distraction. An excess of self-control may cause us to miss out on important life events that arise out of spontaneity and letting ourselves go.

A related problem has to do with the highly contextual nature of human well-being.[3] Many people with exceptional talents that are prized by others still suffer from poor well-being. For example, child prodigies are known for being unhappy, both as children and when they grow into adulthood

(Solomon 2012, 405–476). Their perfectionist tendencies may contribute to their unhappiness since the prodigiously talented can be overly sensitive to even minor imperfections in their performance. It is common for these exceptionally gifted people to commit suicide. Another reality is that natural-born talent is usually not enough to realize success in any given field. Musical prodigies, for example, must engage in constant practice in order to perfect their performance. The need to practice several hours a day can be socially isolating. And a child's talent is not likely to translate into well-being if the child is mistreated in order to develop that talent. There are many stories of parents who are tyrannical in driving their child prodigies toward success (Solomon 2012, 405–476). The various contextual influences on flourishing can derail progress toward well-being (M. Parker 2007). The social attitudes of peers, socioeconomic circumstances, and family life all influence a person's well-being. When these external factors combine with the person's psychological and physical abilities, there is no predicting the outcome for well-being. There is no way of genetically controlling a child's environment or the unknown future.

Savulescu and Kahane could likely respond to these criticisms by pointing out that their Principle of Procreative Beneficence does not require them to claim that parents should maximize any of these (or any particular) capacities. The principle only commits them to claiming that parents should select characteristics that will maximize well-being, whatever those characteristics are. But the problem is with Savulescu and Kahane's deeper presuppositions about what contributes to well-being. The Aristotelian and contextualist insights above show that capacities complement one another in complicated ways, and that no characteristic by itself can act as a determinant of well-being. The simple selection of a set of talents, characteristics, and abilities is not a likely path to well-being for one's child. Savulsecu and Kahane have too much confidence in such a path. Once these presuppositions about what contributes to well-being are set aside, however, it becomes clear that having a disability like Down syndrome can coexist with a state of well-being that is the same as someone without a disability.

My critique of Savulescu and Kahane's Principle of Procreative Beneficence is not meant to show that efforts to further a child's well-being will always be fruitless. But the attempt to have the "best child" with a maximum of valued psychological or physical capacities will be no guarantee of the child's well-being—and might even be counterproductive (see especially M. Parker 2007). Michael J. Sandel proposes an alternative. In his book *The Case against Perfection*, Sandel argues that the best way to promote our children's well-being is through unconditional love, acceptance, and by viewing children as gifts (Sandel 2007). According to Sandel, the urge to select or enhance children so that they maximize valued attributes is a symptom of "hyperparenting," which also manifests as extreme efforts to cultivate athleticism, musicality, and scholastic achievement in one's children. Hyperparenting turns children into instruments of their parents' ambitions. Sandel's account shows that we lose much that is valuable when children are objectified, molded, promoted, and packaged for success. It cannot be good for their well-being.

Hyperparenting

Savulescu and Kahane dispute that there is any connection between the position they advocate and hyperparenting. They claim: "Parents who obsess about their child's well-being and future accomplishment may indeed make their child less rather than more happy or accomplished. But this has nothing to do with the act of selection itself. Selecting the best is not, in this way, self-defeating.... It is subsequent attitudes to the child that may cause such damage. But if so, then this is no real objection to PB [the Principle of Procreative Beneficence]. It is an objection to certain styles of hyperparenting" (Savulescu and Kahane 2009, 284). Savulescu and Kahane seem to make a distinction here between attitudes toward children before birth and attitudes after birth. The urge to control a child's destiny through genetic selection prior to birth they consider to be laudable and fully consistent with acceptance of one's child and a refusal of this urge to control after birth.

In *Choosing Tomorrow's Children*, Stephen Wilkinson (2010, 34) makes a similar distinction. Wilkinson believes that it is a moral requirement to have a "commitment to accepting and loving your child, *once it arrives*" (2010, 34 emphasis original). But he does not believe this norm of acceptance requires refraining from being selective of one's potential offspring before birth. A prospective parent can be willing to reject potential offspring before they are born as long as they do not reject their actual children after they are born. Wilkinson, Savulescu, and Kahane all endorse the norm of acceptance of one's children after birth, however the children turn out, but they see no reason why one should be similarly accepting of one's future offspring before birth. After all, a fetus is different from a child.

It is of course true that the person who is controlling and selective of their offspring before birth will not necessarily become controlling and unaccepting of them once they are born. It is always possible to designate a time "t," and argue that for any purportedly negative attitude demonstrated by a person before t there is no *logical* entailment that the person will demonstrate that attitude after t. But logical necessity is the wrong standard to apply here. It is *empirically* true that personality traits are likely to persist from one time period to the next. Though people can change, such traits tend to persist. Because of this, an attitude toward one's future children in which one is likely to reject a future child (prenatally) because the child exhibits a given characteristic is morally criticizable because such a person will likely display such a rejecting attitude toward his or her children after birth. In addition to the danger that harmful attitudes will persist, morally questionable attitudes are criticizable in themselves. A tendency to reject a child because of his or her disabilities is a morally questionable attitude at any time.

When it comes to disabilities like Down syndrome, as we have seen from the memoirs in chapter 3, it is possible for people to change their attitudes and become more accepting of their children after birth. However, the memoirs also show that when change does happen and parents embrace a norm of acceptance, they are likely to look back and repudiate their previous attitudes of selectivity. They tend to see this embrace of a norm of acceptance as a form of moral growth. So there are two parts to this argument. First, if an attitude of rejection toward one's children is morally suspect after birth, then such an attitude is also suspect before the child's birth, because harmful attitudes tend to persist. Second, to the extent that harmful attitudes do not always persist, and a person adopts an attitude of acceptance toward a child with Down syndrome after birth, this change usually involves retrospective moral criticism of one's previous attitudes. These experiences reveal that legitimate moral criticism can be leveled at the motivation to select and control the characteristics of one's children.

Another line of response to Savulescu and Kahane's claim about hyperparenting is possible. Their own arguments and their views about the well-being of children commit them to an endorsement of hyperparenting. To see this point, consider two assumptions Savulescu and Kahane make about well-being. First, they see well-being as a quantitative concept—as something that can be accumulated and compared to other people's accumulations of well-being. Children are seen as having more or less well-being than other children, even among groups of children in which all are doing well. Savulescu and Kahane argue for a moral obligation to choose children who are "expected to enjoy the most well-being" (Savulescu and Kahane 2009, 274). Second, they see well-being as a competitive concept. Parents are expected to position their children so that they are "more advantaged" than other children to realize this high amount of well-being (Savulescu and Kahane 2009, 276). The pursuit of well-being is a kind of competition and those with greater advantages due to genetics will have a better chance of achieving more well-being. Of course, all of these assumptions about well-being are undefended by Savulescu and Kahane, and there are other ways of understanding this concept. For instance, the concept of well-being could perhaps be better understood as a theoretical category in which people whose lives are going well are included, and within this category there are no differences (or at least no quantifiable differences) as to degree of well-being.[4]

A third point is that Savulescu and Kahane often argue for moral equivalence between measures that promote a child's well-being before and after birth. They argue that actions taken to promote a child's well-being before the child's birth are morally as justifiable as actions that promote well-being after the child is born. Savulescu (2005, 38), for example, states that "If we accept environmental manipulations, by force of consistency we must accept genetic." Manipulating a child's genes in order to give him or her a better life is considered to be similar to "environmental" measures such as feeding the child healthy food, or giving the child a good education. Savulescu and Kahane tend to believe that moral norms of parenthood that apply after birth also apply before birth, and vice versa—(in spite of their pre- and post-birth distinction in the discussion of hyperparenting).

So if there is a moral requirement prior to birth that parents should position their children with advantages in the competition for large amounts of well-being, then by force of consistency Savulescu and Kahane should endorse a similar moral requirement after birth. If there is an obligation to make the "best child" prior to birth, then there seems to be an obligation to turn your child into the best child after birth. In accordance with their assumptions about well-being, doing so would require an accumulation of more well-being than other children in a competitive environment. A parent who sees the world in this way could easily turn into a hyperparent. The difficulty of course is that being a "hyperparent" is likely to undermine the well-being of one's child. This difficulty arises out of the questionable assumptions made about well-being. Perhaps well-being should not be understood as something to be accumulated or won through competition.

Savulescu and Kahane would likely point out that this criticism is simply a result of a "misapplication" of the Principle of Procreative Beneficence (2009, 279) rather than a criticism of the principle itself. If being a hyperparent undermines the well-being of one's children, then the principle would require that one not be a hyperparent. But if this injunction against hyperparenting means that parents should not be selective and attempt to further, at all costs, characteristics in their children thought to contribute to well-being, then Savulescu and Kahane's whole project recommending such measures in the prenatal context seems misguided. Such recommendations might not be misguided if there was a strong asymmetry between ethically justified attitudes toward children prior to their birth and ethically justified attitudes toward children after their birth. But as we have seen, the argument for such an asymmetry appears weak.

Summary: Procreative Beneficence

For the reasons I have outlined, readers need not accept the Principle of Procreative Beneficence, which implies that it would be morally wrong to choose a child with Down syndrome. Savulescu and Kahane's defense of this principle is based on questionable assumptions about two important issues. These authors assume that a child's characteristics further his or her well-being in a fairly straightforward fashion such that choosing the right genetic characteristics will give the child a greater chance at a better life. They also think that well-being can be accumulated so that one person's share of well-being can be compared to another's, and that having a greater chance of accumulating a large share of well-being requires that one have competitive advantages, for example, through one's genetics. These assumptions lead their defense of procreative beneficence into unwanted conclusions. Once these assumptions are shown to be questionable, it is obvious that bringing a child with Down syndrome into the world can be a welcome event, rather than being morally wrong, as the authors would claim.

Reconstructing the P1 Intuition in *From Chance to Choice*

As discussed above, the central intuition of the wrongful disability argument in *From Chance to Choice* (Buchanan et al. 2000) is generated by the P1 thought experiment. If the woman in P1 takes a safe medication and waits just a month to become pregnant, she will have a child without a disability. There is certainty, simplicity, and ease to this decision almost never found in real-life reproductive situations. The decision is a certain exchange of one prospective child who will have a disability for another who will not.

In real life, there is no guarantee that prospective parents will have a nondisabled child the second time around. The second prospective child might have the same prenatally identifiable disability, such as Down syndrome. The fetus might have a disability that is not prenatally identifiable. The child might acquire a disability during childbirth or by an event like an accident or an illness after birth. It might become apparent that the child has a disability of unknown origin sometime after birth. P1 presents an illusion of parental control that does not exist in real life. This illusion of control likely contributes to the intuition that the woman in P1 does something wrong. If prospective parents can easily and certainly prevent their prospective children from having a disability, it seems obvious that it is wrong for them not to.

In contrast, when Down syndrome is suspected through maternal serum screening, pregnant women have to undergo amniocentesis in order to get a diagnosis, and then a second-trimester abortion in order terminate the pregnancy before facing the uncertainty of trying again. The new noninvasive prenatal tests may eliminate some of these difficulties, but they will never eliminate the uncertainty. It is not so easy in real life to avoid having a child with a disability.

The argument in *From Chance to Choice* recognizes the burdens of going through with a prenatal testing regime, of selective abortion, and of trying again for a different child (Buchanan et al. 2000, 250). As I mentioned earlier, the argument excuses prospective parents from the judgment that it is morally wrong to give birth to a child with a disability if the "burdens" of avoiding this are too high. Such excuses are built into principle N, which recognizes "substantial burdens or costs of loss of benefits on themselves or others" as reasons one should not be seen as morally wrong for deciding not to selectively abort (Buchanan et al. 2000, 249). The "excuse" scenarios are regarded as exceptions to the general rule revealed by P1. That is, P1 is regarded as the central thought experiment that generates a reliable indicator of prospective parents' moral duties in reproduction, even though the P1 scenario is totally unrealistic and the exceptions come from real life.

One important disanalogy between P1 and real-life selective abortion decisions is that P1 does not involve abortion. In P1, avoiding the birth of a child with a disability simply involves waiting a month to conceive, and taking a safe medication. Selective abortion is relevantly dissimilar because termination of a pregnancy involves the rejection of a fetus. This element of rejection is notably absent from the situation in P1.

Furthermore, Buchanan and colleagues support the intuition that the woman in P1 acts wrongly by comparing P1 to P2 and P3 (which involve treating a fetus or an infant with medication in order to avoid disability). The woman's actions in P1 might seem wrong because of P1's similarity to P2 and P3. The juxtaposition of P1 to P2 and P3 makes P1 seem like a case of negligence. But again, the comparison with real-life selective abortion situations falls apart. Selective abortion is relevantly dissimilar to P2/P3, because treatment of a fetus or infant for a disability (as in P2/P3) is a totally different kind of intervention than abortion. One implies care and acceptance, while the other implies rejection. The moral obligation to treat a child's disability does not imply that one is obligated to reject a fetus with a disability detected prenatally. These disanalogies are a problem for the wrongful disability argument. Buchanan and colleagues attempt to build an argument about the wrongness of choosing a child with a disability by using a thought experiment (P1) that fails to capture a morally sensitive element of selective abortion decisions.

Many women would regard selective abortion as the rejection of a child.[5] Though the nature of this act would make selective abortion difficult for such women, principle N makes selective abortion after a prenatal diagnosis of a disability morally *obligatory*. The principle states that the actions it recommends are "morally required" (Buchanan et al. 2000, 249). To be sure, the exceptions built into principle N would excuse those who have qualms about abortion from being judged morally wrong. But it is puzzling that the prima facie obligation to selectively abort is derived from a scenario (P1) that cannot account for the special moral difficulties of such a decision. It is odd that the unrealistic scenario of P1 should be regarded as foundational and "authoritative," while typical selective abortion decisions are seen as exceptions to the rule. Perhaps P1 should be regarded as the exceptional case, and typical selective abortion decisions as the more central scenario. If intuitions about real-life selective abortion cases were given more authority, Buchanan, Brock, and colleagues would be less likely to argue that there is a moral obligation to selectively abort when the fetus is found to have a disability.

To illustrate this point, I will present an alternate scenario with more realistic details about a pregnant woman (H) and her partner (J) choosing to have a child with a disability (Down syndrome). I will then ask whether this scenario generates the intuition that the pregnant woman and her partner have done something morally wrong, as the wrongful disability argument would accuse them of doing.

H finds out through prenatal testing that the fetus she is carrying has Down syndrome. She knows selective termination is an option and that she and her partner, J, could try for a child without a disability after terminating. Both H and J believe that parents should unconditionally love their children, and they regard the fetus as their child, though they recognize that others in a similar position might not share their view. They embrace unconditional love as a virtue of parenthood and they want to live up to this ethical standard. The couple also supports the ideal of diversity in their community which they believe comes in many forms: cultural, linguistic, religious, and intellectual. Being pro-choice, H and J have no problem with abortion per se. H does not believe that the abortion procedure in itself would be overly burdensome for her, and there are no foreseeable medical risks or financial costs if the couple did decide to terminate and try again for a nondisabled child. But both H and J believe this would mean losing a child whom they have come to deeply cherish. So they decide to continue the pregnancy, and H gives birth to a beautiful girl who has Down syndrome.

Does this story make you believe that this couple has done something morally wrong? From an ethical standpoint, it seems like an innocuous choice that they have made in keeping with their values and belief systems. But, according to principle N, their choice is either (1) morally wrong or (2) an exception to the principle that such choices are morally wrong. I will consider both alternatives.

If Buchanan and colleagues were to pursue alternative (1), they could use the "opportunity" criterion found in principle N in order to accuse H and J of doing something morally wrong. They have chosen to bring a child with less opportunity into the world when they could have terminated the pregnancy and attempted to have a child without Down syndrome who would therefore have more opportunity. In response to this argument, the couple could point out that there is no reason why the value of opportunity (or "well-being," which, in Buchanan and colleagues' interpretation is constituted in large part by "opportunity") should override other values. In particular, the couple's choice to continue the pregnancy is consistent with their values of unconditional love and their endorsement of diversity as a cultural value. There is no reason why the value of "opportunity" should be weightier than the couple's other values, nor any reason to suppose the couple will not give their child a good life or that their daughter will not bring benefit to the world she enters in so many ways. The wrongful disability argument is no match for respecting parental wishes that arise out of the parents' strong moral ideals and ethical norms.

If, on the other hand, Buchanan and colleagues were to pursue alternative (2), they could excuse the couple from wrongdoing because terminating a wanted pregnancy would have been too burdensome for them. The burden comes from the fact that terminating the pregnancy would require the couple to violate their strongly held beliefs. This response is more defensible than accusing them of moral wrongdoing. The problem, however, is that this kind of excuse would open up a field of exceptions to the wrongful disability argument and principle N that is so large that there would be virtually no grounds for the charge of wrongful disability. As long as a woman who is pregnant with a fetus affected by a disability decides to continue her pregnancy for some strongly felt belief, then her decision is not morally wrong. Only when she makes the decision whimsically or for some trivial reason (e.g., "the delay would interfere with her vacation travel plans"; Buchanan et al. 2000, 244) could the pregnant woman be accused of wrongful disability. But prospective parents do not tend to make reproductive decisions lightly, particularly ones that challenge norms or invite stigma. So the scope for accusing them of wrongful disability—even if we were to accept Buchanan and colleagues' dubious characterizations of "opportunity" and "well-being"—is so narrow as to be virtually nonexistent.

Conclusion

In this chapter, I have addressed a second substantive objection to my position, namely, the argument that it is morally wrong for prospective parents to give birth to a child with Down syndrome or other disability when they could instead selectively terminate and try to have a child without a disability. A version of this argument—known as the "wrongful disability argument—is offered by Buchanan and colleagues in the book *From Chance to Choice* (2000) and elsewhere. I have outlined the wrongful disability argument and its exceptions that excuse prospective parents from moral wrongness when they choose not to terminate a pregnancy affected by a disability.

To rebut the wrongful disability argument, I have shown that the claims Buchanan and colleagues (2000) make about the lower well-being of people with disabilities are questionable. They also propose a narrow understanding of the concept of "opportunity," which focuses on economic competitiveness. Buchanan and colleagues accord opportunity an undue priority in reproductive decisions. More important, however, their wrongful disability argument is based on a supposedly "commonsense" intuition arising from the P1 thought experiment. I have shown how P1 is critically dissimilar to real-life reproductive situations, and I argue that it is implausible to derive a prima facie moral obligation to selectively abort fetuses diagnosed with Down syndrome from this unreliable and questionably relevant thought experiment.

6 Is Selective Termination Morally Wrong?

As we saw in chapter 5, the wrongful disability argument in *From Chance to Choice* (Buchanan et al. 2000) proposes that knowingly bringing a child with Down syndrome into the world is morally wrong. It is possible, however, to support prenatal testing and selective abortion without advancing such a contentious position. For instance, it could be argued that, even though there is nothing morally wrong with continuing a pregnancy affected by Down syndrome, there is also nothing morally wrong with choosing to end an affected pregnancy through selective termination. According to this argument, human fetuses do not have full moral standing, so fetuses themselves are not harmed by ending a pregnancy. Furthermore, according to this argument, people already born who are living with Down syndrome are not affected by the private medical choice to end a pregnancy. Consequently, no harm is done to anyone by selective abortion.

If this reasoning is convincing, then the view that there is nothing morally wrong with selective abortion would be a serious challenge to my thesis that more pregnant women and couples ought to continue pregnancies when they find out that the fetus has Down syndrome. If there is literally no harm done by termination in these circumstances, then the moral "ought" I propose here has no basis. Either choice—termination of a pregnancy or continuation of a pregnancy when carrying a fetus diagnosed with Down syndrome—would be morally innocuous.

In order to counter this line of argument, one must either show that abortion harms fetuses or selective abortion in particular harms people currently living with Down syndrome. As mentioned in chapter 1, my position with regard to abortion is strongly pro-choice. There is little reason to believe that abortion does harm or morally significant wrong to a fetus (see Kaposy 2010b, 2012). I do believe, however, that selective abortion rates of

60–90 percent when a fetus is diagnosed with Down syndrome (Natoli et al. 2012) are a product of harmful attitudes toward people living with Down syndrome. This chapter is devoted to a defense of this claim.

I begin by developing a detailed disability critique of selective termination for Down syndrome. According to this argument, people living with Down syndrome and their supporters may justifiably raise a moral objection against high rates of selective abortion for Down syndrome. There is good reason to believe that biases against people with Down syndrome are a strong motivating factor lying behind these high rates of selective termination. I argue that people with Down syndrome are harmed by bias. Furthermore, because the wider use of noninvasive prenatal testing will likely lead to an increase in selective termination, there is reason to raise a moral objection against making these tests a routine part of prenatal care.

If high rates of selective termination for Down syndrome participate in harmful biased attitudes against people with Down syndrome (as I contend), then the choice of selective termination is not a morally innocuous act. The view that there is nothing wrong with selective termination would then be doubtful. As I will discuss, my position is *not* that every single act of selective termination is morally wrong. The high rates of selective termination in aggregate are objectionable since these rates seem to be influenced by bias against people with Down syndrome. Some decisions made by pregnant women or couples to terminate pregnancies affected by Down syndrome may be justifiable. But because individual acts of selective abortion contribute to the high rates and might well be motivated by biased attitudes, the decision to selectively abort should give pregnant women and couples pause. Because it is the overall trend of selective abortion that is morally questionable, the disability critique of selective abortion for Down syndrome suggests that more prospective parents should choose to give birth to children with Down syndrome.

Overview of the Argument

The disability critique of selective abortion is the position that a moral objection may be raised against the practice of selective termination for disabilities such as Down syndrome.[1]

People living with these disabilities, and those who sympathize with them, have legitimate grounds to raise such an objection. As I discussed in chapter 2, the disability critique of selective abortion has often been equated with the "expressivist objection," according to which the prenatal testing and termination of fetuses with disabilities express negative or offensive messages about people living with disabilities (see, for example, Holm 2008; Parens and Asch 2000; Wendell 1996, 153). Because of these negative messages, prenatal testing and abortion of fetuses with Down syndrome are deemed morally wrong.[2] Various authors have critiqued this argument from a semantic standpoint, arguing that actions such as genetic testing and abortion are "insufficiently rule-governed" (Nelson 2007b, 462), to express any particular message (Nelson 2000; Kittay and Kittay 2000). Although the disability critique has been formulated in expressivist terms, it need not be.

Here is an overview of the nonexpressivist argument I develop in the following sections in support of the disability critique:

1. Bias against people with Down syndrome is a factor explaining the high rate of termination after a prenatal diagnosis of Down syndrome.
2. People living with Down syndrome are harmed by these attitudes of bias, for example through exclusion from informal social relationships.
3. Selective termination for Down syndrome does not itself cause bias, but is a consequence of bias.
4. Even though the high rates of selective termination are a consequence of bias, rather than a cause, one can raise objection to these high rates because we are justified in objecting to actions motivated by harmful bias.

I present arguments that support each of these four points in the next three sections.

Notice that the target of this critique is the high rate of selective termination for Down syndrome. There may be additional or separate reasons for objecting to the routine use of prenatal testing techniques in themselves. Selective termination has certainly been made possible by the availability of diagnostic tests such as amniocentesis. But studies show that many women undergo prenatal testing even though they have no intention of terminating their pregnancies. In one study (Bryant, Green, and Hewison 2010), for example, 67 percent of pregnant women who self-reported as the least likely to terminate an affected pregnancy used maternal serum screening. So the argument developed here as an objection to high rates of selective termination may not in itself be sufficient to justify an objection to a test such as the maternal serum screen. I should note as well some precedents to the argument that I have outlined. There are similar arguments in various articles by Adrienne Asch and David Wasserman (2003, 2005; Wasserman and Asch 2007).

Portions of my argument that bias is an influence on high rates of selective termination (point 1) bolster the well-developed "synecdoche" analysis provided by Asch and Wasserman (2005). I will present Asch and Wasserman's synecdoche analysis in the next section and take up their argument further in chapter 7.

The arguments showing that people with Down syndrome are harmed by attitudes of bias (point 2) draw on ideas about relationships of intimacy advanced by Asch and Wasserman. My arguments here are meant to provide more definitive support for their observation that "harm [is] done to people with impairments when their opportunities in life are constrained by prejudicial attitudes and actions" (Asch and Wasserman 2005, 183).

My analysis of the causal relationship between attitudes of bias against Down syndrome and the high rates of selective termination (point 3) is meant to clarify an issue that is not addressed clearly in Asch and Wasserman's work.

My discussion of the grounds for raising an objection to the high rates of selective termination (point 4) is meant to address a criticism of Asch and Wasserman's position formulated by Jamie Lindemann Nelson (2007a). Asch and Wasserman (2007) themselves do not pursue this line of argument in their response to Nelson.

Bias as a Cause

As we have seen, termination rates after a prenatal diagnosis of Down syndrome are in the range of 60 to 90 percent (Natoli et al. 2012), which means that large majorities of prospective parents do not desire an outcome in which their child has Down syndrome.

In a number of empirical studies (Bryant, Green, and Hewison 2010; Choi, Van Riper, and Thoyre 2012), researchers have asked pregnant and nonpregnant women about what they would do if given a prenatal diagnosis of Down syndrome, and the reasoning behind these hypothetical choices. One study has investigated the reasons that women give for their actual decision to terminate after a prenatal diagnosis of Down syndrome

(Korenromp et al. 2007). In this study, originating out of the Netherlands, the women who had terminated their pregnancies after a prenatal diagnosis of Down syndrome identified a series of motivations, which can be organized into two groups. One group of motivations comprised the beliefs, fears, and desires that the respondents had about themselves or their other family members. More than 60 percent of the women stated that

- they considered the burden of having a child with Down syndrome too great for their other children;
- they considered the burden too heavy for themselves;
- they did not want a disabled child;
- they thought they would become unhappy with having this child (Korenromp et al. 2007).

The other group of motivations comprised beliefs, fears, and desires that the respondents had about their affected fetus, understood as a prospective child. More than 80 percent of the women stated that

- they believed that the child would never be able to function independently;
- they considered the abnormality too severe;
- they considered the burden for the child himself/herself too heavy;
- they worried about the care of the child after their own death (Korenromp et al. 2007).

In one of the studies involving hypothetical decision making, researchers found that beliefs about negative quality of life for parents who have a child with Down syndrome and negative attitudes toward people with Down syndrome, were significantly associated with their intention to terminate an affected pregnancy (Bryant, Green, and Hewison 2010). In this study, the term "negative attitude" had a broad definition and included any emotion such as sadness or pity that could be assigned a negative valence.

The great unmentionable and largely undetected factor in these and other empirical studies is the influence of attitudes of bias against people with Down syndrome in the decision making of prospective parents. No one would be willing to admit that they hold a bias against people with Down syndrome. People are usually able to find an interpretation of their attitudes that does not appear to have the connotation of bias. But bias can have an unconscious influence and its effects can be detected in the motivations that people are willing to acknowledge. Attitudes of bias can be implicit and those holding these attitudes may be unaware of them.

Bias is perhaps closest to the surface in the claim that "I did not want a disabled child"—offered by 63 percent of respondents in the study of women who had terminated their pregnancies after learning the fetus had Down syndrome (Korenromp et al. 2007). This statement expresses a rejection of the disabled prospective child simply because of the child's disability. The disability is a stigmatized trait. In one sense of the word, a stigma is an undesirable characteristic that marks someone as apt for rejection, or as someone who should be denied association. In this case, the stigmatizing trait of disability can change a wanted pregnancy into an unwanted one as soon as it is detected through prenatal testing. There are similar suggestions of bias in the claim that "I thought I would be unhappy with having this child"—offered by 61 percent of respondents in that same study (Korenromp et al. 2007). When this reason is offered, awareness of the single stigmatized trait of disability makes prospective parents unhappy with a previously wanted pregnancy.

Another form of bias can be detected in some of the motivations that relate to the prospective child's life. Bias can be detected in beliefs that a child with Down syndrome will not be able to live independently, that his or her disability will be too severe, or that the burden of life itself will be too great for the child. Instead of seeing the future child as a whole human being who will enjoy all kinds of possibilities for happiness and flourishing, prospective parents adopt a narrow view in which the potential child is viewed exclusively through the lens of a diagnosed disability. As Asch and Wasserman (2005, 181) argue, this kind of motivation for selective termination exemplifies an "uncritical reliance on a stigma-driven inference from a single feature to a whole future life." This is the "synecdoche" argument that testing and termination result from stereotypical generalizations about the potential child based on one piece of information provided by a test. Synecdoche is the literary device in which a part of something represents the whole. It is a form of stereotypical thinking to assume that "this one piece of information suffices to predict whether the experience of raising that child will meet parental expectations" (236). The biased "synecdochical" feature of this reasoning lies in the fixation on the disabled characteristic of the prospective child. The bias leads prospective parents to believe that a child with Down syndrome will necessarily lead a bad life or will be a disappointment because of Down syndrome.

Stereotypes tend to distort perceptions. Some of the common motivations for selective termination reflect inaccurate assumptions about living with Down syndrome or parenting a child with Down syndrome. In the empirical study I have been discussing, 83 percent of respondents who had terminated were motivated by a belief that Down syndrome would be excessively burdensome for the prospective child (Korenromp et al. 2007). In contrast, the study by Brian Skotko, Susan Levine, and Richard Goldstein (2011c; discussed in earlier chapters) reports that 99 percent of people they surveyed who are living with Down syndrome are happy with their lives.

Among respondents who had terminated in the Netherlands study, 73 percent believed that the burden of having a child with Down syndrome would be too great for their other children (Korenromp et al. 2007). Again, in contrast, research involving parents of children with Down syndrome shows that 95 percent of parents with other children say that their children with Down syndrome have good relationships with their nondisabled siblings. More than 90 percent of the nondisabled children themselves say they have feelings of affection for and pride in their siblings with Down syndrome (Skotko, Levine, and Goldstein 2011a).

In the interviews with women who had undergone selective termination, 64 percent had terminated because they felt the burden of bearing a child with Down syndrome was too great for themselves (Korenromp et al. 2007). But research with parents of such children tells a different story. More than 97 percent reported feelings of love and pride in their children with Down syndrome, and only 4 percent expressed regret about having such a child (Skotko, Levine, and Goldstein 2011b). The divergent findings of these studies suggest that perceptions about parenting a child with Down syndrome are distorted by stereotyped ways of thinking and thus also by bias. According to their reading of the evidence, Asch and Wasserman (2005, 175) argue that: "The most that can plausibly be claimed is that being or having a child with a disability is at times different and more difficult than being or having a 'normal' child, and that specific impairments are very unlikely to meet specific parental expectations (e.g., a child with Down syndrome is not likely to become a great mathematician like her mother)."

Table 6.1 summarizes the divergent findings.

Perhaps this line of argument is unfair and the reasoning of prospective parents should be accepted at face value—which is to say that there is no bias against people with Down syndrome in these decisions. One could interpret the empirical studies of people who opt to terminate a pregnancy affected by Down syndrome to mean that women and their partners simply make decisions in accordance with what they perceive to be best for their

Table 6.1
Comparison of perspectives on Down syndrome

	Perspectives of women who have selectively terminated because of a diagnosis of Down syndrome (Korenromp et al. 2007)	Perspectives of people living with Down syndrome, parents of children with Down syndrome (Skotko, Levine, and Goldstein 2011a, 2011b, 2011c)
Quality of life of child with Down syndrome	83%—Down syndrome would be excessively burdensome for the child himself/herself	99% of people with Down syndrome are happy with their lives
Effects on siblings of having a brother or sister with Down syndrome	73%—a child with Down syndrome would be too great a burden for his/her siblings	95% of parents report that children without Down syndrome have good relationships with their siblings who have Down syndrome
Effects on parents of having a child with Down syndrome	64%—having a child with Down syndrome would be too great a burden for myself	97–99% of parents express love and pride toward their children with Down syndrome

families and for themselves. But there is another compelling argument that bias against Down syndrome must have an effect on decision making about termination for Down syndrome.

First of all, there is good evidence that widespread bias against Down syndrome exists. There are some signs that things are getting better, such as the Down syndrome self-advocacy movement. Nonetheless, one still regularly hears people use the words "retard" or "retarded" to signify someone or something that is stupid, broken, incompetent, or disliked. Recent Hollywood movies such as "The Hangover" (2009), "The Change-Up" (2011), and "Ted" (2012) have made people with cognitive disabilities the butt of jokes through the gratuitous use of the word "retard" or plot devices making fun of people with Down syndrome. These days, no Hollywood script would see the light of day if it had similar derogatory material about racial or ethnic groups, yet it still seems acceptable to make fun of people with cognitive disabilities. The currency of the "retard" slur in our culture has given rise to a countermovement to end to use of the "r-word," including a website listing thousands of people who have made public pledges to "eliminate the derogatory use of the r-word" (http://www.r-word.org). Such a movement would not be needed if bias and stigma did not exist.

Furthermore, social science research into the implicit attitudes that people hold toward those with cognitive disabilities is strongly suggestive that negative attitudes are the norm (Enea-Drapeau, Carlier, and Huguet 2012; Hein, Grumm, and Fingerle 2011; Proctor 2012; Robey, Beckley, and Kirschner 2006; Wilson and Scior 2015). In a review article summarizing several studies in this area, Michelle Clare Wilson and Katrina Scior (2014, 315) conclude that all of the studies in their review show "moderate to strong negative implicit attitudes towards individuals with [intellectual disabilities]." The reviewed studies include one that specifically investigates implicit attitudes toward people with Down syndrome (Enea-Drapeau, Carlier, and Huguet 2012). Implicit attitude testing uses methods to determine how strongly people associate pairs of concepts, such as "disabled" and "terrible," "nondisabled" and "pleasant," or vice versa. Such tests can also measure the association between positive and negative concepts and visual images, such as the faces of people who have Down syndrome (Enea-Drapeau, Carlier, and Huguet 2012). The degree to which response times diverge between positive and negative attributions gives an indication of subconscious association between concepts and images (Wilson and Scior 2015).

These implicit attitude findings tend to contrast with findings that measure the *explicit* attitudes that people hold toward those with cognitive disabilities. Explicit attitude studies tend to find that people have very positive explicit attitudes toward this group (Ouellette-Kuntz et al. 2010; Wilson and Scior 2015). However, researchers suspect that explicit attitude research methodologies, such as surveys, can be complicated by the fact that research participants might be influenced by the desire to be viewed favorably by others, and thus may give answers that are more socially acceptable, rather than answering honestly (Ouellette-Kuntz 2010). This problem has prompted the move to other methodologies, such as those used in implicit attitude research. There is also a study using a qualitative methodology showing that explicit positive attitudes toward people with cognitive disabilities can coexist with, or mask, attitudes that are more actively hostile (Coles and Scior 2012). These findings about implicit negative attitudes toward people with cognitive disabilities are consistent with the idea that bias against people with Down syndrome is common as well as unacknowledged or unrecognized.

Second, it seems highly unlikely that widespread bias against stigmatized disabilities like Down syndrome exists but does not affect decisions about carrying a fetus, giving birth to and parenting a child with the stigmatized trait. The high termination rates for fetuses diagnosed with Down syndrome do not indicate that prospective parents are immune to common cultural attitudes toward Down syndrome. Rather, we would expect a high termination rate for fetuses exhibiting a trait that is widely stigmatized in our culture. Of course, there are sure to be exceptions to the norm: individual cases in which women selectively terminate for Down syndrome for reasons innocent of bias. But, looked at from the perspective of the whole population of prospective parents who decline to parent a child with Down syndrome, the pervasive bias against cognitive disabilities must affect many of their decisions.

To be clear, I am not arguing that all decisions to selectively terminate after a prenatal diagnosis of Down syndrome are necessarily motivated by bias. Furthermore, it would be difficult to quantify motivation by bias such that one could argue that most selective termination decisions have this motivation. I have not done this. Instead, I have argued, first, that bias can be detected in the most common reasons offered by women for their selective abortion of a fetus with Down syndrome. Second, I have argued that since bias against people with Down syndrome is common in our culture, it must also be common in selective termination decisions. Neither of these arguments could successfully characterize all individual decisions to selectively terminate as biased decisions. These arguments are effective, however, in showing that bias against people with Down syndrome is an influence—even a significant influence—from a population-wide perspective on the high rates of selective termination. These high rates are the target of the disability critique, as I have formulated it.

The Nature of the Harm

At this point in the argument, it is unclear how selective termination is linked with any recognizable negative effects on anyone. The disability critique is advanced as part of an intramural debate between groups that are pro-choice. Because of these pro-choice commitments, proponents of the disability critique would not allege that selecting against Down syndrome results in wronging or harming any fetuses. According to the pro-choice view shared by those advancing the disability critique, the fetus is not a subject who can be wronged in this way. If the fetus cannot be harmed, that leaves the population of people living with Down syndrome as the group that must be suffering from the negative consequences of the high rates of selective termination (if there is any moral wrong involved).

But defenders of prenatal testing and selective termination have been skeptical of any putative link between these practices and harm or wrong done to people with Down syndrome. Jamie Lindemann Nelson (2007a, 476) points out that, "It will be difficult to provide good evidence for the counterfactual 'Had there been no social practice of prenatal testing and termination, people with disabilities would, other things being equal, have experienced less stigmatization.'" Nelson is right that selective termination in itself does not cause stigma. As I have been arguing, stigma and biased ways of thinking have actually been a primary cause of the push for prenatal testing and the high rates of termination, rather than being an *effect* of these practices. If so, then the common social practice of testing and termination for Down syndrome, and any hardships experienced by people with Down syndrome are not related as cause and effect, but rather both are effects of a common cause—stigma and biased attitudes toward Down syndrome. In a subsequent section, I will explore the basis for objecting to the high rates of selective termination for Down syndrome, given that this practice is not a cause of bias against people with Down syndrome. This section discusses one aspect of the harm experienced by people with Down syndrome resulting from the bias directed at them.

People with cognitive disabilities live within an atmosphere of bias. There are similarities between the effects of this bias on people living with Down syndrome and the way that this bias is manifest in prenatal decision making. The parent-child relationship is among the most intimate relationships in which we may ever be involved. The desire to avoid being a parent of a child with Down syndrome can thus be interpreted as a refusal to be involved in an intimate relationship (i.e., parenting) with a person who has Down syndrome.[3] Similarly, recent Canadian research into the lives of young people with various disabilities reveals a parallel form of this bias. One study has found that more than half of the young people with cognitive disabilities (including Down syndrome) who were interviewed had either no friends or only one friend (Snowdon 2012, 6). The research showed that Canadian communities do a lackluster job of including people with cognitive disabilities in social activities. A longitudinal study in the United Kingdom revealed that, as they get older, children with Down syndrome "begin to have fewer social contacts and friends who are not disabled. By the teenage years many are relatively isolated and become increasingly dependent on the family for social interaction" (Cunningham 1996, 89). An expert on the social geography of people with cognitive disabilities, Deborah S. Metzel (2004, 440) notes that, "Research on the friendships and social lives of retarded citizens has revealed that the majority have social contact primarily with people with mental retardation or other disabilities, with staff associated with service provision, and with family." The key to counteracting the "loneliness and social isolation" (440) resulting from this arrangement is the creation of more inclusive communities. At present, people with cognitive disabilities find it difficult to break into the spheres of intimacy that would constitute inclusion—exemplified by meaningful friendships or involvement in informal social groups.

There may be an apparent contradiction here. On the one hand, I have pointed out that people living with Down syndrome report high life satisfaction (Skotko, Levine, and Goldstein 2011c). On the other hand, there is evidence of their social exclusion. But the apparent contradiction might not be substantive. It is possible to enjoy your life and still think that things could be better for you. The reports of high life satisfaction may be a result of the successful inclusion of people with Down syndrome in family life. A few generations ago when children with Down syndrome were sent to institutions, where they lived in abysmal conditions, they were largely excluded from the family realm. The feature that is often missing these days from the lives of people with Down syndrome is successful inclusion in the wider social life of their communities.

These findings about social exclusion touch on one of the fears that parents of children with Down syndrome have about the future of their children. Parents often express their fear of what will happen if their child with Down syndrome outlives them (see, for example, Bérubé 1996, 85). Family life can provide a network of intimates for a person with a disability. Without a network of friends, siblings, or other sources of social association, the death of one's parents may cast one into a forbidding world without the kind of intimate relationships that typically sustain human lives. Adults with cognitive disabilities whose parents have died may have other caregivers who interact with them and look after their well-being. But designated caregivers provided through public social services or through private arrangements are inadequate substitutes for meaningful relationships between intimates. We are social beings who need friends and confidants and lovers who value us for who we are in ourselves rather than relating to us out of contractual obligation or professional beneficence.

It is often acknowledged that the relationships we have with people who are closest to us—our family members, our closest friends—result in enhanced ethical responsibilities toward these people. For example, I have a stronger responsibility to look out for the well-being of my child than I do for a stranger. Philosophers have described these enhanced moral duties as the morality of "special relations" (McMahan 2002, 218). For many of us, the most valuable aspects of our lives revolve around these relationships of intimacy—whether they derive from intense friendship, relationships with lovers, or from familial ties. Those who are not given this "special relationship" status are owed the baseline of ethical obligations that prevail among strangers. The danger, of course, is when vulnerable disabled persons become strangers to all because they are intimates of no one.

Social exclusion of people with disabilities—exemplified by the friendlessness of young adults with cognitive disabilities—is one of the effects of bias. These harmful effects coexist with official legislative efforts to diminish discrimination against people with cognitive disabilities, such as the Americans with Disabilities Act. For this reason, it is unconvincing to rebut the disability critique by pointing out that "the rise of prenatal screening has coincided with more progressive attitudes toward the inclusion of people with disabilities, as evidenced in the United States by the passage of the Americans with Disabilities Act" (Steinbock 2000, 121).[4] According to this argument, high rates of selective termination cannot be linked with harmful bias because prenatal testing and selective termination have been developed in societies that have enacted progressive legislation and policies that promote the inclusion of people with disabilities. But informal sources of bias exist beyond the reach of the law; they come to light when we realize, for example, that having a disability can be an intensely lonely experience. It would be impossible to legislate that people with Down syndrome be included in meaningful friendships and other intimate relationships. Nonetheless, these relationships are crucial for the well-being of anyone. We suffer when bias prevents our involvement in such relationships. So inclusion of people with cognitive disabilities into our intimate spheres beyond the family will require changes in common social attitudes of bias that exist outside the jurisdiction of legislation.

The Grounds for Objecting to High Rates of Selective Abortion

I have been arguing that the high rates of selective termination for Down syndrome are an effect, not a cause, of attitudes of bias. But if we cannot blame the high rates for the prevalence of bias against people with Down syndrome and their experience of loneliness and isolation, what grounds are there for objecting to these high rates? For one thing, the refusal of intimacy that we can see in decisions to terminate for Down syndrome mirrors the refusal of intimacy that is at the root of the friendlessness and isolation of people with cognitive disabilities. But because the high rates of selective termination are not a cause of bias, there might be a tendency to view this practice as a phenomenon unrelated to any of the harms experienced by people with disabilities. According to this response to the disability critique, we should focus on the real problem—reversing biased attitudes—and leave prenatal testing and selective termination alone. For instance, Nelson argues that: "Struggling against stigma and other forms of disrespect is of the first importance, of course. However, the relationship between these evils and prenatal testing is problematic; who is victimized by any relationship that may exist is unclear and would probably be best addressed by changing features of social life that—unlike prenatal testing—clearly and directly constitute harms and wrongs to people with disabilities." (Nelson 2007a, 477). Nelson concedes the existence of "stigma and other forms of disrespect" against people with cognitive disabilities. In another work, Nelson (2000, 207) recognizes that prenatal testing and selective abortion could be effects of stigma and bias: stating that "the complex of prenatal diagnosis and abortion is at worst the symptom, not the disease." While acknowledging the link with bias, Nelson downplays the significance of this link. For Nelson, the high rates of selective termination are, at most, innocuous by-products of the bias against disabilities, According to this view, "who is victimized by any relationship that may exist is unclear" (Nelson 2007a, 477). It is the fetus that is aborted—fetuses have no standing—and people living with Down syndrome or other disabilities are considered to be untouched by decisions to abort. They are indeed hurt by bias directed against them—the same bias that drives a lot of selective termination—but the harm comes from the bias itself, not from selective termination.

However, the view that selective termination is an innocuous by-product of bias cannot be seriously maintained. Nelson obviously endorses the view that one can legitimately object to biased attitudes against people with disabilities, including Down syndrome. However, her position is to separate cause from effect. According to Nelson, we can legitimately object to the cause (bias against people with disabilities) but not to the effect (selective abortion). However, this position is very odd. There is no problem with objecting to other effects of the bias against Down syndrome. Another effect of this bias is the use of the r-word. One can legitimately raise an objection when someone utters this slur. One can object to the exclusion of people with Down syndrome from educational or social activities. These are further effects of biased attitudes. Nelson suggests that "struggling against stigma and other forms of disrespect is of the first importance" as a way of improving the lives of people with disabilities. But struggling in this way requires objecting to all of the different ways in which bias is manifest. Bias is manifest in the effects of biased attitudes, such as social exclusion, derogatory language, and in this case, high rates of selective abortion. In reference to Nelson's symptom and disease analogy (Nelson 2000, 207), sometimes you cure a disease by treating the symptoms.

In response, Nelson might claim that, to constitute grounds for a legitimate objection, the effects of biased attitudes must be directly harmful or wrongful. According to this position, it is not enough that the features of social life in question are effects of biased attitudes—they must also be harmful or wrongful effects. But I can see no good reason for this distinction. I can legitimately raise an objection to use of the r-word even though it does not harm me directly—and even when it does not harm anyone else directly for that matter. Consider a scenario in which the word is used in the company of nondisabled people who soon forget it. It is still objectionable. What makes the slur objectionable is that it perpetuates a biased attitude that harms other people. Objection to a consequence of biased attitudes held within society is justified because the consequence arises out of morally questionable and harmful attitudes. We may therefore object to high rates of selective abortion on similar grounds. Even though it is not directly harmful or (perhaps) wrongful to anyone, the prevalence of selective abortion after diagnosis of Down syndrome is, in many cases, the manifestation and perpetuation of a common biased attitude. As such, we may legitimately object to this practice. The atmosphere of bias in our society against people with Down syndrome and other disabilities results in their suffering and isolation.

This comparison between the r-word and selective abortion is not meant to make the point that selective abortion expresses a definite message in the way that an utterance of the r-word does. I would not want this argument to be mistaken for a straw-man "expressivist" position. The point is merely that the actions motivated by attitudes of bias are fair game for moral objection regardless of whether those actions are directly harmful or wrongful. What matters is their source in bias that is harmful or wrongful. To give another analogy, a ban on Muslims entering the United States might purport to do no harm to American Muslims. But to the extent that the ban is motivated by Islamophobia, there is reason for American Muslims and others to object to it, as well as to fear the motives underlying the ban.

Future Harm and Selective Termination

I have painted a picture of living with Down syndrome in which there is a risk of social isolation and loneliness, especially after one's parents have died. Prospective parents might think these risks constitute sufficient reason to choose selective termination. They might reason that selective termination would enable them to avoid causing their prospective children harm and distress later in life. I have argued that, in many cases, selective termination is an outcome of biased attitudes that also lead to the exclusion of people living with Down syndrome. Rather than being a reason to refuse selective termination these biases might be understood as a reason to choose it—a significant objection to my position.

But many people without disabilities have trouble "fitting in" for various reasons: shyness, physical, cultural, or sexual differences. We tend not to treat the possibility of social exclusion later in life as a strong reason to terminate a wanted pregnancy when disability is not involved. The reasoning should not be different when disability is involved.[5]

An even stronger response to this objection is to point out that to choose selective termination because of fears about the future isolation of prospective children with Down syndrome is to contribute to, or at least to go along with, the social problem of bias that is the cause of social isolation. Selective termination, when it is motivated by bias, is morally wrong. Even defenders of the permissibility of prenatal testing and selective abortion such as Glover (2006) and Wilkinson (2010) recognize this problem of bias.

But even though choosing selective termination in order to avoid the possible social isolation of one's child is understandable, it is a morally perverse choice. It is morally perverse because the state of affairs that causes social isolation (namely, bias against people with cognitive disabilities) is also a motive for selective termination. Prospective parents who selectively terminate in order to avoid the social isolation of their prospective children behave in the same way that other such parents would if they held the biased attitudes that are the cause of that social isolation.

As I have argued, people living with Down syndrome, who might experience these harms, are the victims of the biased attitudes behind the high rates of selective termination. Choosing to terminate for Down syndrome and thus contribute to these high rates does not help them. But choosing to bring a child with Down syndrome into the world—refusing an attitude of bias—on the other hand, can. Such a choice is a refusal of an attitude of bias. When more families choose to include children with Down syndrome, there are also more self-advocates and natural ambassadors for Down syndrome leading to more welcoming attitudes in our communities toward people with cognitive disabilities. More people like this are needed to counteract the problem of loneliness and isolation. When more prospective parents make this choice, one's child might not have to face the problem of social isolation in the future. Thus the choice to continue or terminate a pregnancy affected by Down syndrome is also a choice about the kind of world we want to live in—one in which bias reigns, or one in which we embrace difference. In order to create a morally ideal world, more prospective parents have to make the morally ideal choice. For those who value tolerance, diversity, and unconditional love of their children, the morally ideal choice is to bring a child with Down syndrome into the world.

Of course, there is always the possibility that even if you, as prospective parents, embrace difference and choose to continue a pregnancy affected by Down syndrome, the world is not changed because so many other prospective parents in the same situation choose to terminate. In such a situation, the risk of isolation later in life remains to be faced by your child. Nonetheless, the findings about the well-being of people living with Down syndrome suggest that, even if this happens, yours is not the wrong choice to be in the minority of prospective parents who welcome a child with Down syndrome into the world. In chapter 8, I explain why it is important even for those of us who currently do not have cognitive disabilities to create communities and institutions that are more welcoming of those of us who do.

Conclusion

As I have been arguing, for any given individual decision to selectively terminate for Down syndrome, it would be difficult to assess whether attitudes of bias are present. But looked at from the perspective of the population as a whole, and considering the persistence of attitudes of bias against Down syndrome within the population, we can infer that bias plays a role in many of these decisions. The disability critique I have articulated above thus targets the high rates of selective termination for Down syndrome for moral objection, rather than any given individual decision to undergo selective termination.

One of the features of early noninvasive prenatal testing that makes it susceptible to the disability critique is that this new testing modality has a closer link to selective termination than maternal serum screening. As mentioned earlier, NIPT is an improvement on maternal serum screening because the newer test can provide fairly accurate diagnostic information, rather than just a ratio of risk, early in pregnancy. And, unlike chorionic villus sampling or amniocentesis, noninvasive prenatal testing carries no risk of miscarriage. Accurate noninvasive testing earlier in gestation makes it easier for women to terminate affected pregnancies. NIPT can shorten the timeline between screening, diagnosis, and selective termination. More pregnant women who are screen positive could likely choose to undergo diagnostic testing since the risk of miscarrying an unaffected pregnancy becomes less of a problem. According to the systematic review by Natoli and colleagues (2012, 149) of termination rates after prenatal diagnosis of Down syndrome, in hospital-based studies "higher termination rates were consistently associated with [diagnosis at] earlier gestational age." It is therefore reasonable to believe (as I have argued in chapter 2) that the widespread use of NIPT will lead to higher rates of selective termination for Down syndrome since this test can lead to an actionable diagnosis earlier in pregnancy.

This chapter provides a rebuttal to the position that there is nothing morally wrong with selective abortion for fetuses with Down syndrome. My argument is that high rates of selective termination for Down syndrome contribute to pervasive attitudes of bias against people with Down syndrome that are harmful to them. The disability critique of selective termination—as I have presented it—delineates the grounds on which we may legitimately object to the high rates of selective abortion after the prenatal diagnosis of Down syndrome. I have provided arguments showing that bias against people with Down syndrome has an influence on many selective abortion decisions. People with Down syndrome live within an atmosphere of bias that can result in their exclusion from meaningful aspects of social life. The harms of social exclusion cannot be blamed on selective abortion or prenatal testing, but any social practice motivated by bias is subject to legitimate moral objection. The decision to undergo prenatal testing and selectively abort a fetus with Down syndrome thus cannot be morally innocuous in all cases. For these reasons, more prospective parents ought to continue such pregnancies and bring children with Down syndrome into the world.

Some readers may not be entirely happy with this moral recommendation since it is not a definitive or "hard" moral "ought" that I am proposing. That is, I am not proposing that it is morally mandatory for pregnant women to choose to give birth to children with Down syndrome when diagnosed prenatally. In light of the argument that I have presented in this chapter, some might think that I should advocate legal or policy measures restricting access to either NIPT or selective termination or both. Some might think that if prospective parents often choose to selectively terminate out of bias against people with Down syndrome, then there are moral reasons for advocating access restrictions that would prevent these biased choices. But my position is that there may be legitimate nonbiased reasons for pregnant women to choose selective termination (I discuss some such reasons in chapter 7). It would be difficult to judge, in any given case, whether the decision to selectively terminate is ultimately the result of biased attitudes. So I believe that the most morally appropriate position is to appeal to the values and humanity of prospective parents and ask them to examine their reasons for considering selective termination, while still enabling pregnant women to make their own choices to continue or terminate their pregnancies. In many cases, I believe these reasons for selective termination do not stand up to rational scrutiny, as I argue throughout this work.

The alternative would be morally worse. If pregnant women themselves are not able to make their own decisions regarding termination, then some other person or group of people would have to be empowered to determine whether a selective abortion decision is justified. The history of abortion in Canada (as an example) shows that when the state takes the decision out of the hands of pregnant women, injustice, abuse, and harm result. From 1969 to 1988, access to abortion in Canada was conditional upon approval by groups of doctors known as a "therapeutic abortion committees" (TACs). This regime for limiting abortion access was ultimately struck down as unconstitutional by the Supreme Court of Canada in *R. v. Morgentaler* ([1988] 1 S.C.R. 30 [Can.]). The therapeutic abortion committees were found to be a serious threat to the "security of the person" of pregnant women, which is protected by the Canadian Charter of Rights and Freedoms. The committees often made decisions that were influenced by religious ideology, with no transparency or accountability. Pregnant women were subjected to the whims of these committees when seeking abortion, without any guarantee of fairness or due process. Many women were forced to give birth to children who were not wanted, or to endanger themselves by seeking abortion or induced miscarriage through dangerous nonmedical means. If we take the likelihood of this kind of abuse into account, it is morally more justifiable for reproductive decisions—such as decisions about selective termination—to remain in the hands of pregnant women themselves.

7 Down Syndrome: Identity and Disability, a Normative Pragmatic Account

There is a striking difference between how medical professionals and advocacy groups describe Down syndrome. Medical professionals tend to see the condition as a genetic disorder that causes disabilities, whereas advocacy groups see it as a form of normal human variation and draw attention away from its status as a cognitive disability. I begin by giving an example of each type of account. I argue that Down syndrome's status as a cognitive disability is attributable, in part, to social norms about the functioning of human bodies and minds. This social contribution to disability is not captured by medicalized views of Down syndrome. Contrary to the medical account of Down syndrome, I argue for a normative pragmatic standard for how we should understand this condition. Our characterizations of people who are living with Down syndrome should reflect ethical standards. We can apply a medicalized view of Down syndrome when doing so is supportive of the well-being of people with this condition, but in other circumstances it may be unnecessary and detrimental to their well-being to focus on this view of their disability in our interactions with them.

Toward the end of this chapter, I apply the normative pragmatic standard I have defended to prenatal decision making about fetuses with Down syndrome. I examine the extent to which a reductionist attitude toward a prospective child with Down syndrome is ethically justified—the sort of attitude described as "synecdoche" by Adrienne Asch and David Wasserman (2005). As noted in chapter 6, synecdoche is the technical name for the rhetorical activity of describing a whole by making reference to one of its parts—of taking the part for the whole. Asch and Wasserman argue that prospective parents engage in synecdochical thinking when they view their prospective child through the restrictive lens of a diagnosis of Down syndrome. In doing so, prospective parents ignore the future child's potential, the contributions the child could make to their lives, and all of the other aspects of character that a child with Down syndrome could embody. In my normative pragmatic account of Down syndrome, I argue that such thinking employs stereotypes and is thus ethically unjustified in most cases. But I also outline different scenarios when prospective parents might be justified in engaging in this thinking as they decide about a prospective child with a diagnosis of Down syndrome. The first part of the chapter is a comparative analysis of how Down syndrome is defined and how people with Down syndrome have been characterized through the use of these definitions. In the second part, I apply this analysis to prenatal characterizations of prospective children with Down syndrome.

Accounts of Down Syndrome

In a 2011 pamphlet that is often shared with new parents, the Canadian Down syndrome Society describes Down syndrome in the following terms:

> Down syndrome is not a disease, disorder, defect or medical condition. It is a naturally occurring chromosomal arrangement. Down syndrome has always existed; it happens in all races, geographic areas, socioeconomic communities and genders. Today, one in every 800 babies born in Canada has Down syndrome. (Canadian Down Syndrome Society 2011)

In this description the emphasis is placed on the naturalness and normality of having a third copy of the twenty-first chromosome. The definition makes the implicit claim that Down syndrome is simply an instance of normal human variation. The claim of "normality" contradicts other common definitions of Down syndrome that characterize the condition as a chromosomal disorder and place an emphasis on medical and cognitive difficulties the child is likely to experience. For example, the Mayo Clinic website offers this definition:

> Down syndrome is a genetic disorder caused when abnormal cell division results in an extra full or partial copy of chromosome 21. This extra genetic material causes the developmental changes and physical features of Down syndrome. Down syndrome varies in severity among individuals, causing lifelong intellectual disability and developmental delays. It's the most common genetic chromosomal disorder and cause of learning disabilities in children. It also commonly causes other medical abnormalities, including heart and gastrointestinal disorders. (Mayo Clinic 2017)

The Mayo Clinic definition places an emphasis on problems, disabilities, and disorders caused by extra material from the twenty-first chromosome.

In its 2011 definition, by contrast, the Canadian Down Syndrome Society denies that this condition is a disorder. The society would agree that it is a disability but might disagree about the cause of the disability. The society would also agree that having this condition makes someone different from others, though this is a difference that should be accepted and not stigmatized. Its insistence on the naturalness and the normality of Down syndrome is a form of opposition to the tendency of nondisabled people to marginalize, exclude, or oppress people with cognitive disabilities. The key claim in the society's definition is that "Down syndrome is not a disease, disorder, defect or medical condition." We can acknowledge that having Down syndrome means being different, but we should not accept that this difference makes a person defective, somehow deficient, or less worthy of respect than other human beings.

The Mayo Clinic definition is a textbook medicalized view of disability. The "genetic disorder" of Down syndrome is caused by an abnormality of genetic material, which is then understood as the cause of disability, developmental problems, and health problems. The cause of disability is situated within the affected individual. It is a genetic cause that can be understood with the tools of science and medicine. The disabilities associated with Down syndrome derive from disordered, abnormal, defective functioning of the individual's body. The body is depicted as functioning in a defective way. One reason to avoid the medicalized view of Down syndrome is the easy inference that Down syndrome is a defect. The chain of causation runs from (a) "extra genetic material," therefore (b) "genetic disorder," resulting in (c) "intellectual disability," "developmental delays," and "other medical abnormalities." The medical definition is a story of a defectively functioning body.

To be fair, a defender of the medicalized view, in making a fact-value distinction, could claim that the medical definition does not mean that individuals with Down syndrome are themselves defective in the normative sense, only that the functioning of their bodies is defective in the factual sense. But statements by prestigious medical organizations like the Mayo Clinic influence how the public understands these conditions. The fact-value distinction might be too subtle for the public. An increase in public stigma is a likely consequence of using a word like "disorder." Furthermore, the fact-value distinction is also questionable in this case since a definition of Down syndrome that implies factual defect and that situates this defect within the genetic material of affected individuals will surely imply that the individuals themselves are defective.[1]

The social model of disability is one of the theories of disability that is frequently used to oppose the medical model. According to the social model, disability results from the social isolation and exclusion imposed upon people with physical or cognitive impairments, rather than from any physical or cognitive deficits of the individual. The social model locates the causes of disability in the social context in which people with disabilities live, rather than in the medically defined characteristics of individual bodies. Disabilities can thus be remedied by making changes to the social and physical environment. A well-worn example is the building that is designed with accessibility features for wheelchair users, who are not disabled when able to navigate such an environment. The social changes designed to counteract a cognitive disability such as Down syndrome would be more complicated to imagine. Measures designed to promote the full inclusion of people with cognitive disabilities—such as policies promoting meaningful employment, good housing, enabling a desired degree of independence, and inclusion in social life—could perhaps show that a social model of cognitive disability is viable.

The social model of disability makes a worthwhile point with regard to cognitive disabilities. Such disabilities should not be seen as traits inherent to individuals that render them defective. Consider an analogy. The causal factor at the heart of the Mayo Clinic's definition of Down syndrome as a genetic disorder is "extra genetic material" from chromosome 21. In most of its manifestations, Down syndrome results from a trisomy (an extra copy) of the twenty-first chromosome. However, having an extra chromosome (a trisomy) does not always result in disability. For example, according to Sharan Goobie and Chitra Prasad (2012), about 1 in 1,000 males is born with an extra Y chromosome, a trisomy of the sex chromosomes known as "XYY." Those with the XYY trisomy are generally tall in stature, but otherwise do not differ in appearance from typical XY males. The condition does not correlate with any health problems and usually goes unnoticed and undetected unless the person having the trisomy undergoes genetic testing to investigate other issues (Goobie and Prasad 2012).[2] From this example, we can see that trisomy in and of itself is no reason to label a condition a disability. If anything, trisomy of the sex chromosomes XYY is an unnoticed difference, a "naturally occurring chromosomal arrangement." So why isn't trisomy 21 as well? Clearly, the symptoms, consequences, or phenotype of a trisomy are more important factors than the trisomy itself in designating a condition a disability.

The cognitive impairments and health issues associated with trisomy 21 are well understood. Down syndrome, unlike XYY, is designated as a disability because of these symptoms. Nonetheless, cognitive impairments and health issues are not experienced by all people having Down syndrome—and are experienced by many people who do not have the syndrome. For the purposes of defining Down syndrome as a disability, we are now a step away from the inherent characteristics of individuals. We are no longer talking about objective characteristics of the body, such as genes and chromosomes. Instead, we are focusing on the functioning of the body, the consequences of trisomy 21—and on how bodily performance is received and understood in social space. The social model of disability has some applicability here. A degree of cognitive impairment would not be regarded as a disability in a society where higher-order reasoning was not required for esteem, social inclusion, or the activities of daily living. I am not sure exactly what such a society would look like, but I think this aspect of the social model is true. A further claim of the social model is that cognitive disabilities such as Down syndrome are disabilities because of the reluctance of society to create communities that accommodate difference and have social inclusion built in.

But the social model of disability has its problems. As Tom Shakespeare (2010) and others have pointed out, the social model founders because it insists upon a strict distinction between the impairment of the individual, on the one hand, versus social oppression, on the other. One problem is that according to the social model, disability is attributable entirely to social oppression. Defenders of the social model acknowledge that people with disabilities have physical or cognitive impairments. But they deny, quite implausibly, that these impairments contribute to the disabilities of these people. The social causes of disability are real, but they exist in complex interaction with an individual's impairments. In Shakespeare's formulation, "disabled by society as well as by their bodies, the social model suggests that people are disabled by society not by their bodies" (Shakespeare 2010, 270). Further argument is probably required to show that physical or cognitive impairment is an essential contributor to disability. But, at this point, it is sufficient to outline the nature of this objection to the social model.

For another problem—and one central to my larger position—the social model's strict distinction between impairment and social oppression is very hard to make. It is virtually impossible to imagine an impairment without reference to some kind of social expectation, norm, or standard. For example, someone who experiences a delay or difficulty in learning how to speak due to a cognitive impairment is regarded as impaired only because our society requires spoken communication for much social interaction. To be sure, a defender of the social model might point out that social changes that accommodate nonverbal communication could succeed in eliminating speech impairment as a disability, but then these social changes would also result in the elimination of speech problems as impairments. Difficulty with spoken communication in a society that does not communicate with speech would not be considered an impairment. Such a difficulty would be an unnoticed difference, like having an extra Y chromosome.

The medical model of disability is deficient because it neglects the social contribution to the experience of having a disability. But the social model is also deficient, shifting the same problems associated with the category of "disability" onto the category of "impairment." The attempt by the social model to give a satisfactory account of "impairment" is likely to run into all the problems associated with the medical model's account of "disability" (Barnes 2016). Social norms and social oppression contribute to the way that some differences are understood as impairments, just as they contribute to the way that some differences are understood as disabilities.

Applying these observations, I would characterize Down syndrome as a disability due to the complex interaction between cognitive differences and health problems that are statistically common among people with the genetic differences characteristic of Down syndrome (such as trisomy 21), and the social norms and expectations of the culture in which they live. Furthermore, there is difficulty with making analytically neat distinctions between things like cognitive differences understood as "impairments" and social norms because impairments always already contain within them a social aspect.[3] Individuals with cognitive differences would not be regarded as impaired if their bodies or their functioning did not fail to meet some social norm or set of norms. Elizabeth Barnes (2016, 37) argues that "society as a whole places normative expectations on bodies—what they can do, how they can do it, and in what amounts. Having a disability is—at least in part—having a body that fails to meet these norms." Without a violation of social norms, a genetic difference like Down syndrome either goes unnoticed as a difference or is understood as a difference without consequence, again like having the XYY trisomy.

I have characterized Down syndrome as a genetic difference that causes many people with this condition to fail to meet social norms or expectations for cognitive function. But this characterization of Down syndrome is, by itself, not fully adequate. I am presupposing an understanding of disability that is overly general. If having a disability means having a trait that fails to meet social norms, or "normal functioning," then this definition will capture things we do not consider to be disabilities (Barnes 2016). For example, being gay, being trans, or being exceptionally talented might involve functioning in a way that is a departure from the norm within a given society, but we don't think these are disabilities. I won't pursue this puzzle of definition any further. It might not be possible to provide a fully satisfactory definition of the concept of disability in the sense of necessary and sufficient conditions for what makes something a disability. Like many concepts in common use, "disability" might not be definable in this sense. Different disabilities might exhibit family resemblances to one another, and to other minority groups, rather than remaining within clear conceptual boundaries. For my purposes, it is enough to point out that the concept of disability as it applies to Down syndrome is, in part, a social construction and a production of social norms and expectations.

A Normative Pragmatic Account of Down Syndrome

If we return to the Canadian Down Syndrome Society and Mayo Clinic definitions of Down syndrome, notice that it is possible for something to be both a "naturally occurring chromosomal arrangement" as well as a "genetic disorder" at the same time. Furthermore, many of the categories I have been discussing can be applied to Down syndrome in varying circumstances—categories like "disability," "impairment," "unnoticed difference," and "inconsequential difference." I would like to defend a normative pragmatic standard for characterizations of Down syndrome. If we acknowledge that Down syndrome as a disability is partly a social construction, then how we understand Down syndrome should be within social control. This is not to say that social constructions are fictive, ephemeral, or imaginary. They are not. Social constructions are real, and they have real effects. But it is possible to organize society in ways that minimize disability, as I have noted. I also think that sometimes it is necessary to regard Down syndrome as a disorder. Sometimes it is necessary to treat health problems or to intervene with speech therapy, occupational therapy, physiotherapy, and so on. In such cases, disordered cognitive function will come to the fore in how we see the person with Down syndrome. When we change social circumstances to accommodate the differences associated with Down syndrome, the disability fades into the background, as we interact with people in their full complexity, rather than simply as people with Down syndrome. At other times, a person's disorderly function must be the focus, and must be what we emphasize as one of their characteristics.

The claim that I am making is about characterizations of people living with Down syndrome. I am no longer addressing the question of how to define Down syndrome, but instead I am discussing the characteristics we attribute to people who have the condition. When we interact with people, we focus on different aspects of their characters or personalities, while other aspects fade into the background and are not the focus of our attention. To give an example, Franklin Delano Roosevelt was both president of the United States and a wheelchair user. Many Americans who lived through the Great Depression and World War II under his leadership never knew he used a wheelchair. The fact of his disability is of comparative irrelevance when considering Roosevelt's contributions to American society and to American history. A biographer of FDR, however, might write passages that emphasize his use of a wheelchair in order to give a sense of the man and the personal challenges he faced. Thus, when considering Roosevelt's personal life, it is appropriate to bring his use of a wheelchair into focus, but when we think of him as architect of the New Deal, this aspect of his personal life fades into the background. The aspects of character that we focus on, ignore, or are unaware of change depending on the social circumstances. In this way, second-person attributions of identity and character are fluid.

We should follow a normative pragmatic approach when we characterize people living with Down syndrome alternatively as having a disability or as being merely different or by creating social circumstances in which it is not necessary to acknowledge that they have Down syndrome at all. These characterizations can each apply in different circumstances. The standard is pragmatic because these characterizations can be applied in a nondefinitive, flexible fashion depending on the context. The standard is normative because ethical considerations should determine the appropriateness of a given characterization. Relevant ethical standards include what the person living with Down syndrome wants, the well-being of the person to be characterized, and the well-being of other persons who have Down syndrome. As the Canadian Down Syndrome Society has noted, people with this condition often would like to be treated as typical people who are not marked as having a genetic condition—or as people who are different but accepted as a normal part of human diversity. In other circumstances, such as therapeutic or educational contexts perhaps, it may contribute to a person's well-being to recognize his or her disability.

An implication of this position is that it is sometimes justified to treat Down syndrome as an unnoticed difference, akin to having the XYY trisomy. Indeed, for some people who have Down syndrome, this happens as a matter of course. Some people do not receive their diagnosis until adulthood and live many years without anyone knowing that they have Down syndrome (International Mosaic Down Syndrome Association 2011). Adult diagnosis is most common for people who have mosaicism of the twenty-first chromosome, which is extra genetic material in some but not all cells of the body. These people might feel that it is better for them if they are not acknowledged to have Down syndrome, particularly if the diagnosis is an incidental or inconsequential finding of a genetic test. Not that this diagnosis should be withheld, but the persons diagnosed would be justified in not sharing the information with others.

I am arguing that ethical considerations should influence questions of identity, particularly ascriptions of disability characteristics. In a way, my position could be seen as a plea to be respectful toward people living with Down syndrome. The recommendation that one should be respectful is pretty commonsensical. However, my position is starkly different from the standard medical metaphysics of Down syndrome. In this standard view, Down syndrome just is the presence of extra genetic material from the twenty-first chromosome, which is a genetic abnormality, which causes disabilities. It might be consistent with the medical view to refrain from using certain terms to describe this condition, but from the medical perspective, how you speak about Down syndrome does not change whether someone has this condition.

The standard medicalized view, I have argued, does not recognize the contribution of social norms to the constitution of disabilities. Whenever social norms contribute to a trait being understood as a disability, the presence or absence of these norms can affect whether the person actually has a disability, which leaves the question of attributions of disabled identity open to normative factors. So I believe my position goes beyond the commonsense proposition that we should speak respectfully to and about people with Down syndrome. I am arguing that whether Down syndrome is a disorder, defect, medical condition, mere difference, unnoticed difference, and so on, are questions that are answerable by considering the ethics of applying these labels to people. I am also arguing that some characterizations are appropriate in some circumstances, but not in others. There are ethical reasons to resist essentialism with regard to Down syndrome.

Many medical conditions have a similar social element that contributes to their status as medical conditions. In psychiatry, there are ongoing debates about whether some conditions are indeed mental illnesses or simply different ways of being. Being gay was once considered a mental illness but is now no longer. The "mad activism" movement and the "antipsychiatry" movement question whether there is such a thing as mental illness and argue that the category of "mental illness" is an illegitimate tool used to oppress and coerce people who are different (LeFrançois, Menzies, and Reaume 2013; Szasz 2010). These discourses are attempts, sometimes using ethical arguments, to move away from a medical understanding of mental illness.

In contrast to mental illness, the treatment of addiction has moved toward medicalization, rather than away from it (Smith 2011). It is now common to understand addiction as a disease, rather than as a lifestyle choice for which a person should be blamed (Smith 2011). The genetics and neuroscience of addiction are replacing the moralized view in which the person living with addiction is someone responsible for his or her bad choices. The medicalization of addiction as a disease is a way of removing some of the stigma of addiction (Smith 2011). In this case, medicalization can be liberating. The demedicalization of mental illness and the medicalization of addiction might seem to be opposites. But, in each case, affected groups use ethical arguments to advocate for how they would like to be viewed—mental illness should be demedicalized because psychiatry is oppressive; addiction should be medicalized because its stigma needs to be removed. The use of ethical standards for how we understand a medicalized group links these two examples. A normative pragmatic approach toward how we understand Down syndrome could be another example of how we might use ethics to influence identity claims about medicalized minority groups.

One potential objection to my proposal has to do with the difficulty of regarding someone with Down syndrome as not having Down syndrome—as

I have suggested can and should be done in some circumstances if doing so supports the well-being of the person with Down syndrome. I acknowledge the challenge since the causes of Down syndrome are genetic, and genes are sometimes taken to be central to identity. People with Down syndrome often have characteristic facial and other physical traits that mark them as having this condition. But there is nothing about cognitive disability, or even congenital disability, that requires us to regard someone's disability as constitutive of their identity. Take dyslexia, for example. For the most part, people having dyslexia are not regarded as disabled in every circumstance of their lives. Their disability comes to the fore as an important characteristic in educational contexts: for instance, when learning how to read. But otherwise, in many other circumstances, a person's dyslexia is not regarded as important, or is overlooked altogether. These responses may be attributable to dyslexia's status as an "invisible disability." Nonetheless, this example shows that it is possible to regard someone's disability as an unimportant trait or as tangential to their identity when doing so is supportive of their well-being. Self-advocates who are living with Down syndrome often plead for others to see them for what they can do, rather than through the reductive prism of their disabilities. Thus, to conclude, a good summary of my position is that we should treat Down syndrome much as we treat dyslexia. Sometimes it is appropriate and beneficial to focus on Down syndrome as a disorder, but, at other times, we should regard people living with Down syndrome in their full complexity as fellow human beings without reducing their identities down to their disability.

Synecdoche

A prenatal diagnosis of Down syndrome in the majority of cases is taken to be a reason for refusing a parenting relationship with a prospective child. I follow Christine Overall (2012, 216) in describing reproduction as the creation of a relationship with a child. She says, "To become the biological parent of a child whom one will raise is to *create a new relationship*" (emphasis original). Selective abortion decisions typically involve a change from the pregnancy being wanted to the pregnancy being unwanted because of the prenatal diagnosis. A woman facing an unplanned and unwanted pregnancy chooses an abortion because she cannot have any child. In contrast, someone who chooses selective abortion decides that she cannot give birth to and parent this specific prospective child diagnosed as having a disability such as Down syndrome, when otherwise she would continue the pregnancy. In this context, the prospective parent refuses a parenting relationship with a prospective child because of the diagnosis. Selective termination is a "not this child" choice, rather than a "not any child" choice (Asch and Wasserman 2005; Saxton 2010). As we have seen in previous chapters, people have many reasons for making this choice. It is nonetheless the case that the choice to selectively abort is also a refusal to enter a parenting relationship with a child who has Down syndrome.

As noted in chapter 6, Asch and Wasserman claim that selective termination is motivated by a reductive understanding of Down syndrome and other disabilities. This allegation of reductionism is their "synecdoche argument." By selectively terminating because of a prenatal diagnosis of Down syndrome, prospective parents refuse a parenting relationship with a prospective child who has Down syndrome simply because of this diagnosis. They take one aspect of a whole human being and make general assumptions on the basis of this one characteristic about the child's life and their life as a family with this child. In the rest of the chapter, I will use either the shorthand "synecdoche" or terms such as "synecdochical thinking," to refer to a perspective that reduces a prospective child's whole character and personality to the diagnosis of Down syndrome so that this one piece of information is considered sufficient to determine whether the prospective parents should enter into a parenting relationship with such a child.

Synecdoche is a characterization of a fetus with Down syndrome to which we can apply the normative pragmatic standard—that is, we can consider whether a synecdochical characterization of a fetus is ethically justified. Of course, the normative pragmatic framework only tells us to draw on ethical considerations, whatever they might be, in determining the appropriateness of attributing different disability-related characteristics to those who have Down syndrome. It does not tell us what ethical considerations in particular are most important, or whether a given characterization is justifiable or not. The first step should therefore be to list the ethical considerations that might apply in this case.

As I have made clear throughout *Choosing Down Syndrome,* I do not believe fetuses are the sort of beings who can be harmed by selective termination. Thus, if a synecdochical characterization of a fetus will lead to selective termination, harm to the fetus cannot be an ethical consideration we use to determine whether the characterization is appropriate. It is legitimate, however, to explore whether a tendency to think "synecdochically" is harmful to others who have Down syndrome. "Harm to others who have Down syndrome" is an ethical consideration that would tend to make synecdochical characterizations of fetuses with Down syndrome inappropriate.

On the other side of things, the parental burden of raising a child with Down syndrome might be one ethical consideration that justifies a synecdochical view of a fetus with Down syndrome. I have pointed out earlier that there is no reason in general to believe that parenting a child with Down syndrome is more difficult or less rewarding than parenting a typical child. But individual prospective parents might have particular circumstances that lead them to believe that parenting a child with Down syndrome would be more burdensome for them than parenting a typical child. In such cases, it might make sense for them to focus on the diagnosis of Down syndrome as the most important piece of information about the fetus on which to base a decision about selective termination. The degree of burden associated with an activity is usually part of the ethical justification for being excused from the activity, or for not having an ethical responsibility to engage in the activity in the first place. In this case, the activity in question is parenting a child with Down syndrome.

Other related ethical considerations might be in play here as well, such as potential effects on other family members such as siblings. Again, virtually all the empirical evidence clearly indicates that children with Down syndrome are loved by their siblings and that having a sibling with Down syndrome is beneficial. However, there might be prospective parents whose particular circumstances might not be captured by the general findings of empirical research.

I will begin by discussing consideration of harm to others who have Down syndrome. In general, we can assume that a tendency to view people living with Down syndrome simply through the prism of their disability is a bad thing for them. If we view someone only as a person-with-Down-syndrome, we do not treat them as a person. Such a reductive perspective ignores their whole existence beyond this single trait, their whole complex personality. What Asch and Wasserman call "synecdoche," when applied to people living with Down syndrome, could also be called "stereotypical thinking," or, more simply, "stereotyping," which is a source of discrimination. People living with Down syndrome suffer from the consequences of such thinking when trying to access employment, citizenship rights, adequate housing, or basic respect.

A decision to selectively terminate because of synecdoche is, however, a decision about entering a particular kind of relationship: a parenting relationship. Someone might argue that a tendency to think this way about becoming a parent of a child with Down syndrome might not extend to viewing other people with Down syndrome in a discriminatory fashion. However, this distinction is implausible. Parenting is one of the most intimate relationships we experience in our lives. Most other types of relationships are less intimate, but the stakes are also lower. There is greater reason to be cautious and careful about those who are tied to us in our closest relationships than there is to be about those with whom we have no relationship at all. I would venture that those who are inclined to think in a stereotyping, synecdochical way about people who are intimates, or candidates to become intimates, are more likely to employ stereotypes about those from the same groups who are merely strangers. When the stakes are lower, there are fewer motives to view strangers as whole, complex individuals with many sides and many different characteristics.

When we go back to the case of prenatal decision making, people who have a tendency to view fetuses diagnosed with Down syndrome through a synecdochical prism would be more likely to view other people living with Down syndrome in the same reductive, stereotyped way, to the detriment of these other people. The harm of ableist discrimination, brought on by a tendency to see people with disabilities reductively, is one reason against synecdoche in prenatal decision making. Thus taking into account the ethical consideration of harm to others, the normative pragmatic framework would suggest that it is inappropriate to adopt a synecdochical perspective when given a prenatal diagnosis that one's fetus has Down syndrome.

The Birth of a Child with Down Syndrome into Poverty

But as I have suggested, there are other ethical considerations that we must take into account to determine the appropriateness of synecdochical prenatal characterizations of prospective children with Down syndrome by prospective parents. The perceived burden of parenting a child with Down syndrome is a legitimate ethical consideration. For example, if a family is in a difficult economic situation, and the increased dependency of a child with Down syndrome will mean that a working parent (usually the mother) will have to take a greater amount of time off work to care for the child, then the arrival of such a child could exacerbate the family's economic plight. In this and other scenarios, the increased time, resources, and attention parents need to devote to a child with Down syndrome could put a family into a situation of difficulty. According to Rayna Rapp's research, familial economic reasons are the most cited reasons for seeking prenatal diagnostic testing and selective termination (Rapp 1999).

An important qualification in such a scenario is that it is not the dependency of the child with Down syndrome per se that is the reason for synecdoche, but the fact that the child with Down syndrome is expected to be more dependent than a typical nondisabled child. The increased dependency of a child with Down syndrome over and above the usual dependency of a typical infant is seen as problematic. Recall that the selective abortion decisions we are examining are decisions about whether to have this particular child diagnosed as having Down syndrome, within the context of a previously wanted pregnancy, rather than a decision about whether to have a child at all. This dynamic means that the "working poor" parents in the scenario above are planning to bring a child into the world, and are thus prepared to take on the total dependency of an infant for an expected period of time.

With this qualification in place, it is clear that the concern about the differential level of dependency associated with having a child who has Down syndrome has limited applicability. In general, families are either able to sustain themselves and flourish with the arrival of a new child, or they are not able to flourish, regardless of whether the child has Down syndrome. In most cases in which the arrival of a child with Down syndrome would threaten the flourishing of a family, the arrival of a typical child would as well. For all intents and purposes, in the first few years of life, nondisabled children are as totally dependent on their parents or caregivers as children with Down syndrome. During this early period, many children with Down syndrome are no more or less dependent than typical children. To illustrate, in her memoir, Kelle Hampton (2012) describes an interaction with her daughter Nella's pediatrician. The pediatrician hands Hampton "a small square of paper" the size of a sticky note with a to-do list for Nella's first three years of life (Hampton 2012, 169–170). The to-do list has four points on it—thyroid screening, a couple of blood tests, an eye exam, and a neck

X-ray. Hampton is surprised that Nella's projected medical care over the first three years of her life can be summarized on a sticky note (Hampton 2012, 169–170). From a medical perspective, there is not much that distinguishes Nella's dependency from that of a typical child. Of course, every child with Down syndrome is different, and some require more care than others. Early intervention programs involving speech therapy, physiotherapy, and occupational therapy can increase the differential amount of care given to children with Down syndrome in the early years of their lives. Many children require surgical interventions early in life. But many other children with Down syndrome are like Nella. The higher level of dependency that would threaten the flourishing of a family that is already expecting a child is not always found in children with Down syndrome.

The difference in dependency level between children with Down syndrome and typical nondisabled children emerges more clearly as the children get older. Children with cognitive disabilities can have greater need for assistance with activities of daily living for a longer period of time than children without such disabilities. Their educational and therapeutic needs may also be greater. It would be a mistake, however, to infer too much about the flourishing of families that include an older child with Down syndrome from a catalog of dependencies. To see the bigger picture of how such families function, we must also look at the contributions made by such children to their families and wider communities. The Canadian Down Syndrome Society's annual heroes campaign provides excellent examples of young adults with Down syndrome who improve the lives of those around them. The 2015 CDSS heroes Megan Allard and Angel Magnussen have both led successful fundraising campaigns for programs that support sick children and people with disabilities in their communities (Canadian Down Syndrome Society 2015). Allard is a leader in the Fusion Inclusion program in her home province of Saskatchewan that advocates for the inclusion of people living with disabilities. Magnussen founded an initiative called "Hugginz by Angel" that sends personalized blankets to sick children in hospitals, and has raised more than $100,000 for children's charities in Canada. Though these young adults have Down syndrome and have care needs that are characteristic of their condition, they are as impressive and inspiring as other exceptional people their age who do not have disabilities. In fact, Allard and Magnussen are probably more inspiring as humanitarians than most other young people.

In the case of low-income families that include a child with Down syndrome, community support and robust publicly funded social programs can help address the child's needs and prevent these needs from threatening the family's economic well-being. The fact that the social context of a child's disability can alleviate the economic impact on a family further restricts the applicability of this line of reasoning about the burden of parenting a child with Down syndrome as a reason for synecdoche in the prenatal period. My position is that it would be heartless to criticize a prospective parent's choice to selectively terminate a pregnancy affected by Down syndrome in a case where the family could not cope economically with the arrival of a child with a disability. But in many other cases, as I have argued, the economic difficulties are such that the arrival of *any* child would have the same effect. And such a scenario could only happen in societies that provide inadequate social and financial support for impoverished and working poor families.

Accordingly, a child, or a prospective child, with Down syndrome should not be seen as a threat to the flourishing of a family. In the case of low-income families, the threat is the more commonplace political scenario of societies that are all too eager to accept the poverty of their fellow citizens, and to further punish them for their poverty. The political and economic policies that have this effect are in fact a threat to any family, regardless of whether they have children with disabilities within them.

Economic hardship caused by difficulties related to caring for a child with a disability like Down syndrome can be an ethical consideration justifying synecdoche, and thus selective termination. But this justification weakens when applied to families that are comparatively well off economically, to families that have access to good social and economic support for children with disabilities, or even to families that are so poor that the arrival of any child would be an economic threat. (If the same justification for termination would apply to a pregnancy unaffected by a disabling condition, then it is not a case of selective termination). As is the case with other arguments in this book, I do not support limiting access to selective termination. I would not be so punitive as to advocate a "means-test" for such access.

There are many factors at play in the burdensomeness justification for synecdoche and selective termination. A family's economic situation and the lack of social and economic support are as much factors that justify the termination of a pregnancy affected by Down syndrome as the fetus's Down syndrome itself. That parenting a child with Down syndrome can be more burdensome than parenting a typical nondisabled child can just as well be attributed to the socioeconomic context of families who find it difficult to raise children who are more dependent than others.

"Bad" Reasons for Synecdoche

There is another set of ethical considerations for synecdochical thinking that must be explored. I will loosely categorize these reasons as "bad" reasons for adopting a reductive synecdochical perspective on a fetus with Down syndrome and wanting to selectively terminate a pregnancy. The reasons are "bad" in the sense that they arise out of parental attitudes toward children of which many people would disapprove. I consider these ethical considerations in recognition of the paradox that, on the one hand, the application of stereotypes is generally harmful to the people stereotyped, while, on the other hand, people who tend to apply stereotypes to others might not make the best parents. It is ethical to take steps to avoid harming future children. So if the reasons prospective parents have for adopting a reductive synecdochical perspective are particularly bad, then perhaps such a perspective is fully justified for them if it leads them to avoid bringing children with Down syndrome they would harm into the world.

For instance, imagine prospective parents who are expecting a child and who have a specific goal in mind for their child. Suppose they are mathematicians and they want to raise their child to be a mathematician like themselves. When prenatal testing reveals that the fetus has Down syndrome and thus would be highly unlikely to become a mathematician, then the couple chooses selective termination. William Ruddick (1998) describes examples like this as "parental self-perpetuation." The prospective parents want their prospective child to be like them, to eventually take on a social identity that they share, and they cannot conceive of any other outcome. According to Ruddick, this attitude toward children is ethically suspect ("bad" in my terminology) because it could lead to undue coercion of a child, and also because it is unrealistic to expect a child to accept the choice of only one identity. When children get older, they regularly rebel against such strictures.

Consider another example: prospective parents choose selective termination because what they most want is a child who will be physically beautiful. The couple does not think that people with Down syndrome meet this standard. There is an obvious difficulty with arguing that such prospective parents should consider continuing their pregnancy and should consider parenting a child who has Down syndrome. One might wonder whether prospective parents with this kind of a synecdochical attitude toward children should end up parenting a vulnerable child with Down syndrome who will suffer because he or she will not live up to parental standards. Bad reasons for synecdoche and selective termination come to seem like better reasons.

I have two observations to make in relation to examples like these. The first is that the birth of a child with Down syndrome is often a transformative event for a parent's value system. I discussed this transformative potential in chapter 3. Many parents of children with Down syndrome testify to the fact that the birth of their child reoriented them to a different sense of what really matters. This new set of values often includes a strong sense of acceptance of the new child, regardless of his or her disability. It is therefore possible that the mathematicians or the shallow parents could change as a result of having a child with Down syndrome and shed their reductive synecdochical perspective. Methods of prenatal testing are not 100 percent reliable, and sometimes parents who would have otherwise chosen selective termination end up with a child who has Down syndrome. Many of the Down syndrome parenting memoirists fall into this category. These parents commonly admit that the attitudes they held before the birth of their child they now disavow and regard as ethically suspect. More generally, many parents are caught off guard by the depth of love that they feel when their child is born, particularly a first child. These emotions can change a person. Even people who embark on procreation with questionable ideas about children may end up being excellent parents. In such scenarios, there may be nothing wrong if a child with Down syndrome ends up in such a family. The benefit to society of such people having children with Down syndrome is the increase in the number of those who value acceptance of people with differences.

My second observation is that the rhetorical undercurrent of these two examples (the mathematicians, the shallow parents) is that people with such attitudes about parenting should not be parents at all. Not only would it be harmful for a child with Down syndrome to end up being parented by the mathematicians or the shallow couple (assuming that they do not undergo a "values transformation")—it would also be potentially harmful for any other child, even one without disabilities. It would be implausible to claim that all prospective parents who would selectively terminate for bad reasons would necessarily undergo a transformation of values if they ended up having a child with Down syndrome. So I cannot argue that all such prospective parents should continue their pregnancies when given a prenatal diagnosis of Down syndrome. But I take this position only because the prospective parents in question—those who are unlikely to undergo a values transformation—probably should not have any children at all.

In summary, I have discussed the avoidance of harm for people living with Down syndrome as well as the burden of parenting a child with Down syndrome as ethical considerations for determining the appropriateness of synecdochical thinking in the prenatal period. I have also explored whether bad reasons for selective termination turn out to be good reasons. Synecdochical thinking is, for the most part, ethically troubling. Someone likely to take on a reductive synecdochical view of a prospective child would have a greater tendency to apply ableist stereotypes to others who are living with Down syndrome. But there are some situations in which such a view might be warranted, for example, when parenting a child with Down syndrome would threaten the economic well-being of a family as compared to parenting a typical child. There are important limitations to this burdensomeness argument, as well as to the situations in which "bad" reasons motivate synecdochical thinking.

Conclusion: Dependency as Part of the Human Condition

I have been discussing the perceived burden of parenting a child with Down syndrome as a rationale for synecdochical thinking and selective termination. When determining the ethical strength of a claim that parenting a child with Down syndrome would be burdensome, the particular details of a family's situation matter. Prospective parents might be able to provide the care needed by a child with Down syndrome even though they believe otherwise. They might underestimate the amount of support available to them—particularly if they are unfamiliar with social programs for people with disabilities and their families. They might overestimate the level of dependency of people with Down syndrome. A generalized fear of dependency might lead to the mistaken assumption that children with Down syndrome are too difficult to care for, or that their care will negatively affect their families.

A fear of having to raise a highly dependent child might be at the root of many decisions to refuse to parent a prospective child diagnosed as having Down syndrome. People with physical and cognitive needs that make them dependent have long been marginalized in Western culture and political discourse. Going back to the Enlightenment, according to Kant, personhood and intrinsic value are owed to those who have the capacity for autonomy. In social contract theory, legitimate political power is thought to originate from voluntary agreements between free rational men in a state of nature. In this tradition, human beings deserve respect and recognition of their intrinsic value by virtue of their capacities for autonomy, freedom, and rationality. These are the capacities of the independent person, however, not those of a dependent person. By valorizing the capacities of independence, the standing of dependent people—such as those with cognitive disabilities—becomes marginal.

Our ethical obligations toward independent autonomous persons are made necessary by their valorized capacities. Imposing the same standard on dependent persons makes our ethical obligations toward them seem derivative or optional. In contrast to the intrinsically worthy independent person, the person who is dependent on others becomes the object of charity. Nancy Fraser and Linda Gordon (1994) argue that, in Western political discourse, especially in the United States, the concept of charity is put to use when dominant contractual notions of human relationships fail to explain some of our political or ethical obligations. But charity has become stigmatized in political discourse. Charity is understood as a "pure, unilateral gift, on which the recipient had no claim and for which the donor had no obligation.... In the modern conception of charity, the giver got moral credit while the taker was increasingly stigmatized" (Fraser and Gordon 1994, 101). In conservative political thought, noncontractual nonreciprocal social programs such as welfare are seen as handouts, as getting "something for nothing," an unjustifiable reward for laziness. People receiving social assistance are seen as "takers" rather than "makers." In this social context, when the state furthers the interests of dependent people out of charitable obligation, those who are dependent are seen as a drain on social resources. Within this political history of dependency, it is not surprising that many people have a horror of dependency and are reluctant to parent a child they know will be more dependent than other children.

The problem with this understanding of dependency is that it fails to acknowledge the degree to which we are all dependent on one another, disabled and nondisabled alike. The distinction between those who are independent and those who are dependent is untenable. For one thing, independence is a temporary condition in human life. We are all born into extreme dependency at the beginning of life. Then, throughout our youth and adolescence, the nature of our dependency changes, but the fact of it does not. Eva Feder Kittay (1999, 83) explains that "In complex societies nearly two decades are required to train individuals to be 'fully cooperating members of society,' and in all societies approximately ten childhood years are spent in nearly total dependence on an adult." Throughout adulthood and into old age many experience the temporariness of independence. As Martha Nussbaum (2006, 101) points out,

> the way we think about the needs of children and adults with impairments and disabilities is not a special department of life, easily cordoned off from the "average case."... As the life span increases, the relative independence that many people sometimes enjoy looks more and more like a temporary condition, a phase of life that we move into gradually and all too quickly begin to leave. Even in our prime, many of us encounter shorter or longer periods of extreme dependency on others—after surgery or a severe injury, or during a period of depression or acute mental stress. Although a theoretical analysis may attempt to distinguish phases of a "normal" life from lifelong impairment, the distinction in real life is hard to draw, and is becoming harder all the time.

It is normal and average for us to be dependent on others for significant care needs for much of our lives. It is not just the "other" who is dependent—the child with the cognitive disability, for instance. We are all dependent at different moments in our lives. This fact is hard for many of us to accept because of the value we place on our ability to control our lives and futures (Garland-Thomson 2012). But this power of control is fragile and fleeting, if it exists at all. No one is independent for the whole of a life.

Furthermore, the lived experience of independence is so mixed up with various forms of dependency that a distinction between the two is almost meaningless. For example, dependency is often seen as a threat to economic opportunity. But the realization of economic opportunity is itself a state of almost total dependency on others: in the form of economic and educational institutions, laws protecting private property and inheritance, noncorrupt governmental practices, kinship relations and "connections," stable currencies, international trade agreements, and so on. The Nobel

Prize–winning economist Herbert Simon (2000) argues that at least 90 percent of personal income in a wealthy society like the United States can be attributed to "social capital," which he defines as stored knowledge shared within a society—such as technological knowledge, the rule of law, and organizational and governmental skills—as well as "participation in kinship and other privileged social relations" (Simon 2000). Reliance on networks, relationships, and infrastructure is a form of dependency that capitalizes on investments made by those in generations that have come before us, and on the stored tangible and intangible forms of capital among our contemporaries. Virtually any human pursuit is similarly reliant on other people.

Some may acknowledge the *inter*dependency of economic activity, while maintaining that the agents involved in such relationships are nonetheless independent actors. Kittay (1999, xii) refers to this position as "the pretense that we are *independent*—that the cooperation between persons that some insist is *inter*dependence is simply the mutual (often voluntary) cooperation between essentially independent persons" (emphasis original). According to this position, there are different types of people who are differently dependent. An independent person may enter into a dependent relationship with another independent agent, but this situation is different from the dependency of the child with a cognitive disability, who is not an independent agent. But considering what it takes to function as a supposedly "independent" person, it is impossible to isolate the independent aspect of the person's functioning and separate it out from the various dependencies at play in any activity. These dependencies are multifaceted. As Kittay and Nussbaum point out, the "independent" person can only develop out of the extreme dependency of infancy and childhood. We carry this developmental history along with us throughout our lives since the experiences of childhood have a way of giving us advantages and creating disadvantages (by virtue of the quality of care we received or lack thereof) later in life. Aside from this developmental dependency, all of our activities are enabled by the social circumstances, resources, networks, and entrenched norms of our current situated and lived experience. "Independence" as a concept only makes sense within such a context. Most of the social circumstance that we benefit from are not the result of voluntary cooperation, but are instead accidents of birth or other situated forms of good fortune.

The illusion of independence is sustained against a background of dependency. At best, the difference between the hidden dependency of the paradigmatically "independent" and the dependency of the child with a cognitive disability is a difference of degree, or a difference in the kind of support that is needed by each. The fact of dependency is the same in each case. Thus the notion that we are independent is largely illusory. Our fear of dependency and the stigma we attach to it are responses that do not take into account the normality of dependency.

At the beginning of this chapter, I sought to explore the claim—implicit in the Canadian Down Syndrome Society's definition of Down syndrome—that Down syndrome is a difference that ought to be accepted as an instance of normal human variation. This chapter shows that in many ways people with Down syndrome are relevantly similar to people who do not have disabilities. We are all dependent on one another. We all require care throughout our lives. We all need to be valued in order for our lives to go well. The different ways we characterize people with Down syndrome often turn out to be based on mistaken assumptions, biases, unexamined stereotypes, and a failure to acknowledge the social contribution to cognitive disability. Some of these characterizations can be found in the medical discourse about Down syndrome. All of these characterizations affect prenatal decisions about whether to enter a parenting relationship with a prospective child who has Down syndrome. Rather than taking on unexamined stereotypes, Down syndrome could be seen as an instance of normal human variation. If more of us characterized Down syndrome in this way, the demand for selective termination after a prenatal diagnosis would diminish.

8 Conclusion: The Negative Influence of Capitalism on Reproductive Selectivity

This final chapter offers one further argument for bringing a child with Down syndrome into the world. This argument is meant to directly address my readers and asks you to examine your values and your beliefs about children. In previous chapters, I have argued that there is a pervasive bias against people with Down syndrome in our culture. This bias is manifested in various ways, such as the common use of offensive slurs and hurtful jokes directed at people with Down syndrome and other cognitive disabilities. In this chapter, I probe beneath the surface of the bias to explore a more deep-seated reason why people might not want to have a child with Down syndrome. I draw together several observations made in earlier chapters in order to form them into a coherent argument.

In this chapter, I show that the demands and imperatives of the economic system in which we live influence many to believe that raising a child with Down syndrome is undesirable. According to my analysis, many people have internalized the values of the capitalist economic system, which marginalizes people who find it challenging to contribute to the profit-making enterprise. Because of the difficulty a child with Down syndrome may have in fitting into this economic system, prospective parents might choose to terminate an affected pregnancy. To advance my analysis, I discuss the economic motives in play in the development and marketing of prenatal tests, as well as the economic factors that influence physicians to offer novel tests such as NIPT to pregnant women. I also present some of the economic factors that might influence a pregnant woman's decision to undergo prenatal testing and to choose selective termination.

In pointing out these factors, I would like to show that when we focus on the difficulties associated with raising a child who has Down syndrome, we ought to consider not just the subject of our focus—the child with Down syndrome—but also the frame, or the social context that can make this life difficult. In presenting such an analysis of the social context, I suggest that the problem many prospective parents are facing is not the prospective child diagnosed with Down syndrome, but the values of our society, that impede our ability to make humane and generous choices. The child is not the problem. Rather the problem consists in the social values that present the child as a problem. These social values are highly influenced by the economic organization of our lives. This chapter is a way of drawing attention to these values. When we stop for a moment and examine these values and the economic forces from which they arise, it is worth asking whether we ultimately endorse these values. As I argue below, I do not believe that many readers would ultimately make such an endorsement. I suggest that the choice to test for and terminate pregnancies affected by Down syndrome is influenced at a deep level by a set of values that many would disavow if only they became aware of them. This chapter is a way of drawing attention to these values. The alternative is to refuse this influence and instead choose to bring a child with Down syndrome into the world.

The Marketing of Noninvasive Prenatal Testing

The development of NIPT has been driven by a profit motive. In the research and marketing of these tests, biotech companies such as Sequenom, Verinata Health, Ariosa Diagnostics, and Natera have sought to tap into the prenatal testing market, estimated in 2013 to be worth $1.3 billion a year in the United States alone (Agarwal et al. 2013). In 2014, Sequenom reported revenue of $151 million (*GenomeWeb News* 2015). Marsha Saxton has called the development of these tests "a new 'bright hope' of capitalist health care and the national economy" (Saxton 2010, 130). Saxton argues that the development of new prenatal tests "exploits the culture's fear of disability and makes huge profits for the biotech industry" (Saxton 2010, 128). The common fear and bias against disability are certainly factors driving the greater demand for prenatal testing. But the ability of NIPT to capture a preexisting market is only part of the story.

As I argued in chapter 2, NIPT is perfectly suited to create a new market for prenatal testing among pregnant women who would refuse previous testing modalities— maternal serum screening and amniocentesis, for instance—because of their inaccuracy, or the risk of inducing miscarriage associated with previous tests. NIPT is not just successful in capturing the demand of pregnant women who would previously have accessed prenatal testing in spite of the inaccuracies and the risks. Part of the profitability of NIPT is due to the creation of this new market among other women who would have been skeptical of previous tests. To say that NIPT creates a new market is another way of saying that the introduction of these new tests has changed people's behavior. Many women who would have refused testing in the past will likely agree to prenatal testing now. This change of behavior reflects how the capitalist profit motive has intervened into decision making and shaped the experience of reproduction.

The medical profession has been co-opted into the creation of this new market for prenatal testing and has contributed to the profitability of NIPT. Physicians can be held legally liable if they do not offer prenatal testing to their pregnant patients (Pioro, Mykitiuk, and Hewison 2008; Ossorio 2000). The standard of care according to clinical practice guidelines in many jurisdictions is to offer prenatal testing to all pregnant women (Pioro, Mykitiuk, and Hewison 2008). A failure to do so can be deemed a negligent breach of the standard of care. If a child is born with a disability like Down syndrome after the physician failed to offer prenatal testing, and the parents can successfully demonstrate that they would have terminated the pregnancy had they known the prenatal diagnosis, then the physician could be subject to damages for "wrongful birth" (Pioro, Mykitiuk, and Hewison 2008; Ossorio 2000). Thus physicians have both a legal and an economic motive for offering prenatal tests such as NIPT to all pregnant women (Murdoch et al. 2017). The historian Ilana Löwy (2014) has traced out how earlier forms of prenatal testing became widely adopted, in part, because of the fear of litigation. With such motives in play, physicians become effective agents in the promotion of products marketed by prenatal testing companies.

An unanswered question at the moment is whether physicians have an obligation to offer pregnant women NIPT in particular, rather than just maternal serum screening followed by amniocentesis, in order to comply with the standard of care. Though this issue may not have been tested in court at present, fear of litigation can certainly motivate physicians to routinely offer NIPT to all pregnant women, and this routinization can lead to the emergence of a legal obligation. As Pilar Ossorio (2000, 312) explains, the standard of care among physicians is a herd concept—it is defined by the prevailing actions of all others within the group. The more common a practice becomes, the more likely the practice will be regarded by the courts as the standard of care. The practice of "defensive medicine" motivated by the fear of litigation can cause a practice—such as the offering of NIPT to all pregnant women—to become common. "The more tests physicians order to prevent liability, the more likely it is that they will create a legal duty to offer these tests, regardless of whether testing is otherwise well advised" (Ossorio 2000, 312). "Defensive medicine" results in ordering more tests, even if these tests are truly unnecessary (Bishop and Pesko 2015). In the United States, most doctors admit that they practice defensive medicine (Bishop, Federman, and Keyhani 2010). To be sure, many physicians are motivated to offer NIPT to pregnant women for other reasons: because autonomy is an important value in health care, or because it is regarded as good medical practice to give patients choices. But whatever other motivations that physicians have, there is also an underlying legal and economic motivation to offer NIPT.

Furthermore, medical professionals are also effective agents in the promotion of NIPT products because an offer of a test by such a professional can easily be seen as an endorsement by a knowledgeable and authoritative figure. As Mark Pioro, Roxanne Mikitiuk, and Jeff Nisker (2008, 1028) point out, "the availability of prenatal screening, and the fact that it is being offered by a physician, carries normative implications about the desirability of prenatal screening" (see also Lippman 2004, 403). For someone who may not know much about Down syndrome, the offer of a prenatal test for such a condition by a doctor could clearly be seen as a warning that Down syndrome is necessarily something that must be avoided (Saxton 2000). After all, we usually agree to undergo medical tests to identify diseases and conditions we would rather not have. One does not usually undertake diagnostic tests to identify characteristics that are innocuous or beneficial to one's well-being. For many people, being offered tests is a sign that the physician is providing good-quality care (Bryant, Green, and Hewison 2010). Thus the authority of medical professionals and their ingrained fear of legal liability make physicians well suited to assist biotech firms in their pursuit of profits through the marketing of NIPT.

As Abby Lippman has observed: "'Needs' for prenatal diagnosis are being created simultaneously with refinements and extensions of testing techniques themselves" (Lippman 2004, 403). These needs are created by authoritative recommendations by medical professionals, as well as by biotech firms that create new markets for prenatal testing through the development of new tests. Economic motivations lie behind all of this activity.

Economic Influences on Reproductive Choices

The influence of the capitalist economic system can be also discerned in the beliefs and decisions of prospective parents. Some of these beliefs are given voice in *From Chance to Choice* (Buchanan et al. 2000). I have argued in chapter 5, that this book by Buchanan and colleagues can be read as an economic justification for prenatal testing and selective abortion. I argue that these authors find "opportunity" in a capitalist market system to be the dominant value that should inform selective abortion decisions. They contend that a child who has a disability like Down syndrome will be at a disadvantage in our competitive economic system and thus will have lower comparative well-being than a nondisabled child. Because of this comparison, they argue prospective parents should selectively abort if given a prenatal diagnosis of Down syndrome, and should instead try for a child who will not have a disability. The authors of *From Chance to Choice* argue that prospective parents can cause "wrongful disability" if they instead choose to bring a child with a disability into the world. The wrongfulness of this choice derives from the child's comparative diminished economic opportunity.

In chapter 7, I also discuss the economic circumstances that can make raising a child with a disability seem overwhelming. Many women who were interviewed by Rayna Rapp (1999) in her research into the social context of choosing amniocentesis expressed their concerns about economic vulnerability as a motive for undergoing such tests. The stories presented in Rapp's research "poignantly illustrate how close to the edge many parents feel when they imagine the juggling of work and family obligations should disability enter an already tight domestic economy" (Ginsburg and Rapp 2010, 243). Poverty can be a factor in such decisions, as well as underfunded or nonexistent social programs for affordable housing, a lack of early intervention programs, a lack of affordable health care, and poor access to necessary allied health care such as physiotherapy, occupational therapy, and speech therapy. Education for a child with Down syndrome also requires a public commitment to fund specialized teacher training and individualized classroom assistance. Low-income families will not have access to these supports without publicly funded social programs. Raising a child with Down syndrome could seem like a prodigious challenge without this support. Laissez-faire capitalism and efforts to decrease the size of government can put families, especially low-income or working poor families, into this situation. According to the rhetoric of "small government," low taxes are more important than the social programs families need to raise children with disabilities (Bérubé 1996).

An example of this tension comes from the election campaign in the Canadian province of Ontario in 2014. During this campaign, Progressive Conservative[1] leader Tim Hudak promised to cut 100,000 public service jobs if he was elected (*CBC News* 2014). Given the number of people employed by the provincial government and the magnitude of the proposed cuts, it was inevitable that many teachers and teaching assistants would lose their jobs. Hudak himself admitted this (Maharaj 2014). The reduction in the number of teaching assistants would have been disastrous for children with Down syndrome and for their schools. In order to be included in regular classrooms, most children with Down syndrome require additional classroom assistance. With the crucial support of teaching assistants, the children's education can be a story of success and inclusion, but without it, education can be challenging for both the children and their teachers. Fortunately for the people of Ontario, Hudak was defeated in the election, though job-cutting, tax-cutting, small-government conservatives are in charge in many other jurisdictions. The supposed difficulty of parenting children with Down syndrome can often be traced to the political and economic environment in which families raise their children, rather than to any inherent difficulty caused by the disability itself.

Another example of economic influence can be found in the value systems of many prospective parents. My examination of Down syndrome parent memoirs in chapter 3 shows that many such authors admit that they were preoccupied with middle-class identity and values before the birth of their children. Their memoirs commonly evoke their worries that a child with Down syndrome would find it difficult to reach middle-class milestones of a good life—including going to college, having a wedding, and buying a house. Many of these milestones are synonymous with economic prosperity. Within the context of these values, a significant aspect of a "good life" is having enough money to move through this set of meaningful life events. In these stories, we can see the subtle influence of the economic system on our understanding of what constitutes a good life. If these accounts are true to life, many prospective parents attempt to discern the future well-being of a prospective child by estimating the child's ability to follow the conventional middle-class course of a prosperous life. Moneymaking is considered a measure of well-being. This perspective on childhood is understandable given the political climate in many Western nations. When government programs and the social safety net are slashed, private wealth can be necessary for survival. Prosperity and the things that it can buy (university, wedding, real estate) can be a buffer against the ruin of a family. It is little wonder that we valorize prosperity. As we can see, the values that drive the economic system can easily become internalized so that raising a child with Down syndrome can seem undesirable to prospective parents because of perceived difficulties that the child would have in living up to expectations about what constitutes a good life. The common middle-class view of the good life is itself influenced by the capitalist market system.

The Technological Reconstitution of Pregnancy

The availability of increasingly accurate and user-friendly prenatal genetic testing technologies changes our relationship to the fetus. This change is a further way in which economic concerns have exerted a profound influence over reproductive decisions. Bruce Jennings (2000, 126) refers to "the reality-constituting power of the technology itself" in his description of how prenatal testing can change the way pregnancy is understood. Previously in human history, pregnancy was a mysterious process and the fetus was hidden within the pregnant woman's body. The child's ultimate characteristics were determined as though they were a matter of fate. With the element of chance diminished by the availability of prenatal testing and the means for selecting for or against certain characteristics, the mysterious aspect of pregnancy diminishes as well. Though these technologies are meant to empower choice, their introduction and the changes they have wrought have occurred beyond the sphere of voluntary choice of prospective parents. Expectant mothers and their partners can choose to forgo prenatal testing, but the "burden of choice" itself cannot be avoided once the technology is available (Sandel 2007, 89). That is, prospective parents can refuse testing, but they cannot refuse to make a choice one way or the other—to test or not to test. Before such tests existed, pregnancy could be seen as an area of life in which fate was the rule—in the sense that the prospective child's characteristics were not subject to choice. After prenatal testing became available, this perspective on pregnancy became no longer legitimate, since these technologies have put the ability to choose our child's genetic characteristics increasingly within our grasp.

In opposition to this ability to exercise choice, Michael Sandel writes about how a child welcomed into the world should be seen as a gift. He claims that this perspective leaves us open to the "unbidden" aspects of life. "To appreciate children as gifts," Sandel (2007, 45) tells us, "is to accept them as they come." In his argument supporting this view, Sandel sets up an opposition between openness to the unbidden, and a contrary attitude, the impulse to exert mastery over reproduction.[2] He refers to the latter as the "Promethean impulse"(89). According to Sandel, openness to the unbidden in human life is necessary to maintain valued moral attributes like humility, solidarity, and responsiveness to the needs of vulnerable others (Sandel 2007). These attributes require recognition that we lack control over much that is important in our lives. For Sandel, the view that children should be seen as gifts, rather than as objects of control, and his endorsement of the fateful aspects of life, are part of a larger concern that valuable moral attitudes are in danger when we try to exert technological domination over areas of our lives previously outside of our control.

One argument supporting this position advances the following points: (1) in order to be responsive to the vulnerability of others, we must recognize our own vulnerability; and (2) the "Promethean impulse" leads us to believe less in our own vulnerability. When we are convinced of our ability to control and master, we have less motivation to respond to the needs of vulnerable others. The ability to identify with those who need our help is often a potent motivation to provide aid. We are far more likely to help out an identified victim of a calamity than a faceless "statistical" victim (Daniels 2012). But we are less inclined to help when we can't put ourselves in the shoes of the other. When we do not see ourselves in the vulnerability of others, we often do not respond.

Sandel's concern is that the more we exert mastery over our lives, through reproductive technology, for example, the less open we will be to recognizing our own vulnerability—and thus we will be less likely to identify with others who are vulnerable. If we believe that our children's talents, health, and abilities are contingent products of the natural lottery, we will tend to believe that their well-being owes much to good fortune. Subject to the same tides of good or ill fortune, the needy child could just as well be my child. But if we believe that our children are armored against the contingencies of fate through our prudent use of reproductive technologies, we have less motive to identify the vulnerable child as just like our child. In such a scenario, our ability to respond to the needs of others is diminished. Those who believe that they themselves are solely responsible for their success are likely to ignore the needs of others. Such people are also likely to blame the weak for their weakness. For instance, the rich "self-made" man is liable to blame the poor for their inability to pull themselves up out of poverty.

Noninvasive prenatal testing as a technology of control can be subjected to Sandel's critique. In contrast to an attitude of openness to the unbidden, technologies such as NIPT enable an attitude of mastery—they make the fetus knowable, and more subject to intervention and control. The mystery of pregnancy becomes dispelled by the reality-shaping technology that allows us to read the fetal genome at the earliest stages of development. The child is no longer a gift, unchosen and unpredictable, but becomes instead like an artifact made-to-order (Lippman 2004). If Sandel is right, this loss of a sense of the unbidden in pregnancy is ethically troubling because it is a symptom of larger cultural changes that result in the erosion of valuable moral attitudes such as the ability to respond to the vulnerability of others.

The commercial nature of NIPT can be subject to this critique as well. With the availability of NIPT, the control of the fetus is accomplished through the purchase of a product. Marsha Saxton (2010, 128) writes about the "lure of consumerism" that motivates prospective parents to seek to control the outcome of pregnancy via prenatal testing and selective termination. Faye Ginsburg and Rayna Rapp (2010, 241) note that "consumer capitalism is shaping the experience of reproduction." Pregnancy has become a site of consumer transactions and capitalist profit making. Instead of mystery and ritual, we have commerce. As Sandel suggests, the ability to purchase control over a pregnancy has ominous implications. As a cultural trend, not only does an excess of control and mastery in our lives leave us less responsive to the inevitable vulnerabilities of others—it also leaves others less responsive to our own vulnerabilities. Even when we are convinced of our own mastery over the unbidden, we never have total mastery. We remain vulnerable to some extent, in some areas of our lives. Our mortality and embodied nature make it unavoidable that we are subject to illness, disease, or disability at some point in our lives. For these reasons, we benefit from the responsiveness of others to our needs, sustained by an openness to the unbidden. Pregnancy and parenting are spheres of our lives in which we have an opportunity to preserve a welcoming attitude to the "unbidden," the unexpected. But doing so requires us to resist the consumerist impulse to exert control over reproduction.

An Appeal to Your Values

In summary, we can see the influence of profit making, capitalism, and economic concerns from two different sides of the development and use of prenatal tests. On the side of the biotech industry and the medical profession, the profit motive is paramount in the development and marketing of NIPT. On the side of the consumer, economic concerns drive many prenatal testing and selective abortion decisions. As I argued in chapter 2, a likely result of the promotion and greater use of prenatal testing modalities like NIPT will be ever greater numbers of selective terminations for Down syndrome. At a deeper level, the result of these economic forces on the two sides of the clinical relationship—offering and consenting to prenatal testing—is an elimination of difference. Over time, it is possible that fewer people with Down syndrome will be born. With the expansion of NIPT to test for many other identifiable genetic differences, fewer people with these other differences will be born as well. The elimination of difference is equivalently the creation of a kind of uniformity or standardization among our children. The development of new prenatal tests and the internalized values of economic prosperity conspire to create a certain kind of child. This child is one who is able to follow the arc of a conventional profitable life. This is the child idealized in *From Chance to Choice*—one who maximizes economic opportunity within a capitalist market system. This arrangement is perfect for the creation of people who grow up to become cogs in the capitalist market system, and people who can perpetuate this system.

The purpose of telling this story is to ask the reader whether he or she endorses these values. If the reader does not, there is an opportunity to make another choice. It is possible to refuse prenatal testing. If one undertakes prenatal testing and is told that the fetus has Down syndrome, it is possible to continue the pregnancy rather than terminate. I am guessing that many people who would consider selective termination for Down syndrome do not ultimately endorse the values I have articulated above. To endorse such values, one would have to view the capitalist market system as valuable in its own right—apart from its ability to improve our lives as compared to other possible economic systems. The position that the capitalist market system is valuable in its own right is the position of free market–loving fiscal conservatives. In the current alignment of political values, fiscal conservatives often identify as antiabortion social conservatives as well. This alignment exists in the Republican Party in the United States, in the Conservative Party in Canada, and in conservative parties in other countries as well. Insofar as fiscal conservatives tend also to be against abortion (given this political alignment), they are not people who would consider selective abortion. Pro-choice liberals, on the other hand, typically have no moral objection to abortion, and thus would comprise a likely market for NIPT and selective abortion. But liberals are typically much more wary of the kind of economic policies and influences that are in play in the account I am giving of the development and greater use of NIPT. Few liberals would endorse the capitalist market system as a good in itself. Few liberals would endorse the elimination of human differences that cannot be readily accommodated within the norms of capitalism. To be sure, this depiction of political allegiances is a simplification of the views that people actually hold. But it serves to illustrate common positions and the fact that people who typically support abortion rights might not ultimately support the values that underlie the prenatal testing industry and selective abortion for disability.

If you are a pro-choice liberal who would consider using prenatal testing with the intent of selective termination, you might realize that you do not ultimately support the values I have articulated in the account above. To support these values—as a short form I have labeled them "fiscal conservative"—one must have a worldview in which what matters most is achievement and competition—ultimately economic competition—and an urge to have power and dominate others. We lose a great deal when we view life primarily as a competition. When we strive for mastery and lose the mysterious and unbidden aspects of our lives, we become different people, and the change may not be very good for us. To be responsive to the vulnerability of others we must recognize our own fragility and our own dependence on others. Vulnerability and dependency are our common lot. Without this insight, there can be a tendency to view vulnerability as blameworthy or as inherent in the person instead of as an unavoidable circumstance. As Ian Brown (2009, 181) suggests: "A world where there were only masters of the universe would be like Sparta. It would not be a kind country. It would be a cruel place."

The critique of capitalism I am presenting in this chapter borrows concepts from Marxist philosophers such as Antonio Gramsci. Gramsci and later theorists on the political left developed the idea of "cultural hegemony"—which is the position that dominant ideologies in a culture are self-serving products of the ruling capitalist class (Gramsci 1971; Said 1979). Ideas, beliefs, and values that are accepted as unquestioned norms, or as "common sense," are actually created out of the imposition of power. These manifestations of dominant ideologies are so persuasive that even those who are most victimized by them may not realize their contingent nature and may accept this partial worldview generated by capitalism as the obvious truth. In this chapter, I am defending the position that the belief that it is ethically justified or even morally required to selectively terminate a pregnancy affected by Down syndrome is a product of capitalist hegemony. The decisions to offer prenatal testing, to undergo prenatal testing, and to terminate if given a diagnosis of Down syndrome, appear obvious and commonplace when we are in the grips of the particular market-oriented worldview of our culture. But, as Gramsci (1971) tells us, we can question the influence of hegemony and we can expose its workings. These are, in part, my goals in this chapter.

Though I acknowledge the influence of Marxist thought, my position is also consistent with other critiques of capitalism originating from other quarters. For instance, Catholic social thought has a long and distinguished history of objecting to neoliberal and laissez-faire capitalism. In 1986, the United States Catholic Bishops issued a Pastoral Letter entitled "Economic Justice for All" setting out the commitments of Catholicism to support social programs such as welfare, good public education, and health care. The bishops reason that human dignity can be protected only in a community, and that the community has a responsibility to protect and support the poor and the vulnerable. In his recent encyclical letter "Laudato Si'," Pope Francis (2015) is highly critical of capitalist economic activity that destroys the natural environment, ushers in climate change, creates social inequality, and threatens the well-being of the poor. The vision of society presented in these teachings is one in which the influence of capitalist ideology over reproductive decision making would be diminished—one in which prospective parents would not be forced by economic circumstances to refuse to parent a child with a prenatally diagnosed disability.

In a somewhat similar vein, Karl Marx and Friedrich Engels suggest that it is possible to create a society that is not dominated by a single economic imperative, but instead embraces a diversity of human lives, purposes, and relationships. In a capitalist society,

> each man has a particular, exclusive sphere of activity, which is forced upon him and from which he cannot escape. He is a hunter, a fisherman, a herdsman, or a critic, and must remain so if he does not want to lose his means of livelihood; while in communist society, where nobody has one exclusive sphere of activity but each can become accomplished in any branch he wishes, society … makes it possible for me to do one thing today and another tomorrow, to hunt in the morning, fish in the afternoon, rear cattle in the evening, criticise after dinner, just as I have a mind [to], without ever becoming hunter, fisherman, herdsman or critic (Marx and Engels [1845] 2004, 53).

While I am not insisting on the need for proletarian revolution to enable us to realize a social or economic system that would allow us to flourish, recognizing the influence of the competitive market system over our life decisions is the first step toward avoiding this influence. Humans are not exclusively economic agents. We are also members of families, of cultures, nations, and religions, with interests and needs that extend beyond economic agency. Our reproductive decisions should not be determined by economic imperatives.

Bringing a child with Down syndrome into the world is a way of refusing the values of a dominant economic and ideological system. It is an endorsement of different values, and a way of living a different life. In making this choice we also recognize that we need to build communities and societies that enable all of us to thrive. If we live our lives according to values that we actually endorse, rather than by values that we would ultimately renounce, we can begin building these communities.

Objections

Consistency with Reproductive Autonomy

The argument I have presented in this chapter is an appeal to consistency. I argue that if prospective parents would reject the perpetuation of the capitalist market system as a good in itself, then they should consider bringing a child with Down syndrome into the world. I make this inference because many of the reasons that support selective termination when given a diagnosis of Down syndrome have their roots in a system of values that promotes the perpetuation of the capitalist enterprise as a good in itself.

One way of responding to an argument that advances consistency as a virtue is to downplay the importance of consistency in the issue at hand. Some might say that inconsistency is not so wrong in this case. Prospective parents can choose selective termination if given a prenatal diagnosis of Down syndrome and work against the excesses of capitalism in other ways without being morally bad people. According to this objection, it might be too demanding to expect people, even those who decry the influence of capitalist values on our everyday decisions, to be completely consistent. They can be excused for being inconsistent when complete consistency is overly demanding and they act consistently with their beliefs in other ways.

Of course, holding consistent beliefs and values is not all that difficult. But consistently carrying them out—acting on them—can be. In this sense, the challenge is in parenting a child with Down syndrome. As I have argued throughout this book, this challenge is overstated. Parenting a child with Down syndrome is similar to parenting a child without a disability, and being such a parent has its own benefits that are not always realized by parents of exclusively nondisabled children. For prospective parents with reasonable economic security who have a stable family life there is little difficulty being consistent, or at least the difficulties are most often exaggerated by those who do not know what it is like to parent a child with Down syndrome.

One of the main arguments that supports the use of prenatal testing and selective abortion is the reproductive autonomy argument. According to this argument, if a pregnant woman autonomously chooses to use prenatal testing and likewise makes an autonomous choice to selectively terminate a pregnancy affected by Down syndrome, then we must respect these choices. In the field of bioethics and in the wider culture, the ethical norm of respecting autonomy is so strong that the burden of proof is on those who would prevent autonomous choice. To make a case against respecting autonomy, the choice in question must pose significant harm to the chooser or to others. On these grounds, some might argue that my consistency argument above fails because the ethical principle of respecting autonomous choice in reproduction requires that we respect choices that people make even when these choices might be inconsistent with their values. The problem of inconsistency is easily overcome by the stronger requirement to respect autonomy. The pregnant woman herself is the one who determines whether parenting a child with Down syndrome would be overly burdensome.

As has been obvious throughout this book, I fully endorse this perspective. My position is pro-choice, and I agree that pregnant women are the only people who have the morally legitimate authority to exercise control over their bodies. But I have two responses. The first is that, even though the pregnant woman is the sole person who should be empowered to make her own reproductive choices, there is nothing untoward about offering respectful and rational arguments in favor of one choice (bringing a child with Down syndrome into the world) over other choices, such as selective termination. I make this point in the introductory chapter.

My second response is that my consistency argument draws upon a deeper understanding of autonomy that might not be apparent in the usual formulation of the principle of respect for autonomy. According to one understanding of this concept, the obligation to respect autonomous choice consists simply in a norm of noninterference. Health care providers can respect autonomous choice simply by providing unbiased information, by acting in a noncoercive fashion, and by letting pregnant women choose for themselves. As Victoria Seavilleklein (2009, 72) notes, according to this understanding of autonomy, "women's choice is thereby restricted to their ability to accept or decline a particular option that is offered to them." According to another understanding, however, autonomy is the ability to make choices in concert with one's most deeply held values. On this understanding of autonomy, acting in a way that is consistent with one's values is the essence of autonomy. It is often pointed out that the concept of autonomy derives from the ancient Greek word for "self-governance" (*autonomos*—"having its own laws"). To govern oneself is to make choices that derive from the beliefs and values with which one most strongly identifies. A choice that is inconsistent with one's values could then be seen as a failure of autonomy.

Many philosophers add a "relational" or contextual component to this understanding of autonomy (e.g., Sherwin 1998; Mackenzie and Stoljar 2000; Seavilleklein 2009). Our choices, values, and beliefs are shaped by the interconnected relationships of power in which we find ourselves. The relational approach to autonomy asks us to take into account the influence of various forms of oppression over decision making, particularly the influence of oppression on decisions made by women. The "noninterference" understanding of autonomy "allows no room for reflection on the *practice* that is making those particular choices available or on other contextual influences outside the clinic that may not qualify as coercive or substantially controlling but may nevertheless have a significant impact on women's decision-making" (Seavilleklein 2009, 72; emphasis original). These practices and these contextual influences can influence people to make choices that are inconsistent with their deeply held values. The exercise of autonomy is a set of skills that can be developed or neglected, supported or undermined. The most extreme forms of oppression can clearly compromise the skills needed for the exercise of autonomy. For instance, physical and sexual abuse can have profound psychological effects on self-confidence, feelings of self-worth, and self-trust (McLeod and Sherwin 2000). To act autonomously, one must believe that one's choices can be effective and that these choices matter. The feeling of empowerment as a necessary condition of being autonomous can be attenuated or missing in those who live in oppressed circumstances—for example, as a result of abuse.

Less extreme contextual factors can negatively influence our ability to make choices that are in keeping with our most strongly held values. As I argue earlier in this chapter, medical professionals, in offering their pregnant patients prenatal tests, lend their implicit endorsement to these tests. Because of the authority and prestige of these professionals, it takes a degree of knowledge and self-confidence to be able to critically appraise the offer of a test. Without this knowledge and self-confidence, the test may be uncritically accepted as obviously beneficial, and the condition it tests for may be uncritically viewed as uniformly bad for the prospective child and for the prospective parent. In the background, however, the physician may be offering the test as an economically motivated standard of care, while the prospective parent may strongly endorse values of acceptance of difference, diversity, and unconditional love of her children. The power dynamics of the encounter, gaps in knowledge about disabilities, and external forces can influence a prospective parent to make decisions that are ultimately out of keeping with her own values. The "noninterference" understanding of autonomy does not capture these dynamics.

In this chapter, I have shown how capitalist economic influences constitute the context of decision making about prenatal testing and selective abortion for Down syndrome. I have suggested that if people were aware of these influences, many would ultimately disavow them. This economic context is not always apparent in reproductive decision making. For people who would disavow these influences, a lack of awareness of the economic context can result in decisions that are made nonautonomously—according to a relational understanding of the concept of autonomy. A realization that our choices and needs have been constructed by the economic system in which we live can be the first step toward reclaiming autonomous choice. As I have been arguing, a truly autonomous choice is possible. A choice to bring a child with Down syndrome into the world—contrary to all the influences that push people to do otherwise—can be such a choice. While pregnant women should always be the decision makers empowered to make choices about their own reproduction, we should also empower them to make choices that are in keeping with what they most value.

The Evils of Capitalism and Selective Termination

I have described capitalism as a system that excludes and burdens those of us who do not conform to its requirements. Some might think that living in such a system would be harmful for someone who has Down syndrome and that selective termination would thus be justified as a way of preventing people with Down syndrome from experiencing such harms. This objection is similar to one I countered in chapter 6.

As I have pointed out throughout this work, my pro-choice position is that pregnant women should not be prevented from making the choice to selectively terminate. But, in this case, I think that choosing to selectively terminate because of fears about economic-related harms experienced by a future child who will have Down syndrome would be counterproductive from an ethical perspective. The choice between continuing or terminating a pregnancy after a diagnosis of Down syndrome is a choice between acquiescing to a system that is not good for us, versus trying to change this system. Rather than preventing the birth of people who might be harmed by an unjust system, one could try to change the system itself, or alter it so that our communities are made more welcoming. Of course, prospective parents could also do both, but the choice to bring a child with Down syndrome into the world is definitely more consistent with creating a welcoming community.

Even if noninvasive prenatal testing and selective abortion were eventually successful in eliminating all people with Down syndrome or other genetically caused forms of cognitive disability from the population, there would still be other people with cognitive disabilities. There are nongenetic prenatal causes of disability. People acquire cognitive disabilities during birth, and after birth as a result of accidents, injuries, or infections. Others become cognitively disabled as a result of disease or aging. Genetic intervention and selective abortion will not eliminate these vulnerabilities from the population. If the capitalist system is bad for people with cognitive disabilities, it is better to alter the system so that it becomes more welcoming, rather than trying to eliminate cognitive disability.

Furthermore, those of us who do not currently have a cognitive disability could join their ranks at any moment of our lives. Vulnerability is a fundamental condition of embodiment. Noting "the frailty and brevity and precariousness of human existence," sociologist Tom Shakespeare (2014, 109) believes we should adopt a position of "pessimistic materialism" toward human embodiment. It is in our own interests, not just in the interests of those who currently have cognitive disabilities, for our communities to be welcoming of vulnerability. Selective termination does not advance these interests arising out of our vulnerability, but is rather the sort of reaction encouraged by an economic system that excludes differences. We need institutions and personal relationships that allow people with cognitive disabilities to thrive. Publicly funded high-quality inclusive special education programs, low-cost accessible housing arrangements that promote independence, governmental strategies that get people with cognitive disabilities into the workforce, savings programs with state contributions that allow parents to secure the financial future of their children, publicly funded high-quality health care, measures for including people with cognitive disabilities in civil society and for reducing bias against people who are different—these are the kinds of support that should be in place to counter the exclusionary economic imperatives of capitalism. There is a good chance that coordinated activities like these will be crucial for our own well-being as well.

From this perspective, a choice to selectively terminate would be counterproductive, and contrary to the spirit of openness to difference that is needed to build communities that promote flourishing. For those who have values that are critical of the economic imperatives of capitalism, the choice to selectively terminate because of fears about the evils of capitalism would be a perverse choice. Doing so would be to accede to the influence of capitalist values. A morally ideal choice would be to choose in accordance with your own values about the worth of every human being regardless of ability. A choice to bring a child with Down syndrome into the world would also do a lot of good, since choices like these create natural advocates and ambassadors for Down syndrome within families who can realize the promise of truly inclusive communities.

Afterword

When Aaron was three years of age, he needed heart surgery. At birth, an ultrasound revealed an atrial-septal defect (ASD), which is a small hole in the septum that acts as a wall between the two upper chambers of the heart. We were told that most ASDs close by themselves as the child grows but that there was a small chance Aaron's would require surgical correction in a few years. It turned out that Aaron's atrial-septal defect fell into that small percentage of ASDs that do not heal on their own. As we found out later his heart condition might not have been caused by Down syndrome. Many children without disabilities have ASDs that require surgical correction.

By the time he was three, Aaron had gone through a few bouts of pneumonia, including one for which he was hospitalized. A heart condition like an ASD can result in increased risk of pneumonia. When our cardiologist measured Aaron's ASD and found that it wasn't shrinking, she offered us the option of surgery. We live in Newfoundland on the eastern edge of the North American continent, but we opted to bring Aaron to Toronto for surgery at the Hospital for Sick Children because of the stature of this world-renowned institution, and because of our family ties to the Toronto area, where Jan and I both grew up.

The doctors at the Hospital for Sick Children first tried to close Aaron's ASD noninvasively, using a sophisticated technique that involves inserting a catheter and threading it through the blood vessels of his body into his heart. Though Aaron's doctors discovered very quickly that he was not suitable for this procedure, it still involved preparing Aaron for surgery. We learned that the experience of handing your child over to the surgeons, even for a noninvasive procedure, can be traumatic. On the morning of surgery, Jan brought Aaron into the operating room and laid him down on the bed among the doctors and nurses. Only one parent was allowed into the room. The procedure involved giving Aaron anesthesia, but not until Jan had left. She had to leave him, frightened and crying and alone with strangers in these strange surroundings, knowing that he was about to undergo surgery, with all of the risks operating on the heart entails. In Aaron's short life so far, aside from his birth, this experience was the most elemental. (We had to experience it twice, since the noninvasive procedure was unsuccessful, and he underwent successful open-heart surgery a few weeks later.) The degree of trust involved in bringing Aaron in for surgery was profound. These doctors and nurses were essentially strangers to us, so we could scarcely place our trust in them as individuals. Rather we trusted their professional designation, and the fact that they were practicing in a highly respected hospital. In fact, placing Aaron into their hands in the midst of his fear and ours was more like an act of blind faith than of trust.

When my mother was a child, her younger sister, Donna, died at the age of two. The family story told by my grandmother was that Donna had a "hole in her heart." In pictures, Donna is a happy-looking child who is small for her age. She died in 1953, when the surgery and intensive care technologies available for Aaron's care simply did not exist. Donna's doctors were waiting for her to get older and stronger before risking an operation—but they ran out of time. Growing up, I never heard my grandparents speak much about Donna's life: she was not a usual topic of conversation with grandchildren like me. This is understandable. Nonetheless, I always saw her life as part of the explanation for the large gap in age (eight years) between my mother and my aunt Patricia. Had Donna not died, the age differences between the sisters would not have been out of the ordinary.

My grandmother Evelyn Hunter died in 2006, just over fifty years after Donna's passing. At the funeral, the pastor described a conversation she had with my grandmother a few months before her death. My grandmother had come to her for pastoral care counseling from this trusted United Church minister and had wanted to talk about Donna. Fifty years after her baby daughter had died, my grandmother was still thinking about her and was still looking for help in order to come to grips with Donna's death. My grandmother was an elderly woman who was very happy, and who had lived a very active life traveling and camping with my grandfather just a few years earlier. She had outlived her days of raising small children by several decades. But, right up to her final days, my grandmother saw herself as a mother—the mother of a sick and dying child. This story told at her funeral gave me the sense that the death of her child was always close in her thoughts and memories. As a child, I never had a sense of how important Donna's death had been in the lives of my grandparents, though now it seems completely obvious. The death of a child would have to be an event that shapes the rest of your life and remains with you for the rest of your days.

While we were in Toronto for Aaron's heart surgery, my thoughts often returned to my grandmother and Donna. Aaron's heart defect was similar to Donna's. From conversations with family members, I suspect Donna had a ventricular septal defect (VSD), which is a hole in the division between the two lower chambers of the heart. With Aaron, we were taking on the risk of heart surgery, which seemed like an unfathomable and unnaturally daring undertaking. I thought about the possibility that we could lose Aaron, and wondered whether we had been foolish to welcome the risk of surgery. The doctors and nurses were going to give him powerful drugs to render him unconscious, then cut open his tiny little chest, and somehow keep him alive while they patched up a hole inside his heart. Then they were going to start his heart up again (or hope it starts) and reverse the process of cutting him open by stitching him back together. The horror of considering what heart surgery entailed was matched only by the banal confidence of those who were caring for him. Aaron's cardiologist at the hospital called the surgery "a piece of cake." Though unfathomable to me, pediatric heart surgery is currently at a level of technical sophistication that repairing an atrial-septal defect is considered relatively routine. It might have been routine for them, but it seemed extraordinary to me as Aaron's father.

The cardiologist was right to be so confident. Except for a brief and quickly remedied drop in Aaron's blood pressure, the procedure went well. When we went to see him after surgery, he was unconscious, attached to various machines, and was being cared for by a friendly and trustworthy nurse. I was reassured by her calm and the feeling she conveyed that caring for Aaron was a job she could do very well. Jan had worked for years herself as a nurse in intensive care units, so she knew what to expect when we walked into the unit to see Aaron. My own experience with intensive care was limited to a few clinical ethics consultations with adult patients. To prepare me before we had even left for Toronto, Jan had asked other parents in our local Down syndrome society for photos of their kids in intensive care after their heart surgeries. One image was especially helpful. It showed a little girl we had known from the society after her heart surgery, a few years prior. In the digital version of the photo, all of the tubes, lines, and wires running to and from her little body were labeled. Jan took some time to explain how all this technology worked.

In intensive care, Aaron looked much like the little girl in the photo, who had made a quick and full recovery. He had drains coming from his chest, a catheter running into his diaper, wires for a temporary pacemaker that entered his chest cavity, and various monitors for heart rate and oxygen saturation. His arms and legs were strapped to splints to prevent him from pulling out his lines. When he woke up, Aaron was not entirely happy with the splints or his immobility. But he wasn't in pain or overly upset. Miraculously (in my mind), Aaron spent only two days in the intensive care unit, and then part of another day on a ward. We stayed with him in the hospital for most of our time in Toronto, and his big sister came to visit on the last day. Aaron was discharged home on the third day with only acetaminophen and ibuprofen for pain management, which surprised me, but, as I later learned, this is standard treatment for pediatric patients recovering from heart surgery.

Aaron's recovery at Jan's parents' house just outside of Toronto was difficult and stressful for about a week. He did not want to eat, crawl, walk around, or even look at anyone except Jan. He just wanted to be held and to nurse from his mother. But after about a week, Aaron was eating again and more engaged. He quickly returned to his passion of watching Elmo in YouTube videos. Aaron's heart surgery was completely successful. His heart is now fixed—and functioning normally. We had always had challenges with Aaron's appetite before his surgery. After the surgery, however, he had a more reliable appetite, and became more open to a broader range of foods. He began to gain more weight, and friends and family pointed out that he was getting taller. We suspect that his heart condition was a cause not only of his bouts of pneumonia, but also of his more frequent low-level nasal congestion caused by the various viruses and bacteria that circulate among children. The nasal congestion made him disinclined to eat. With the ASD fixed, much of this problem cleared up, and Aaron was thriving.

The kind of care given to Aaron was routinely denied to infants with Down syndrome as recently as the 1980s in some parts of North America. In the preface and in chapter 2, I discussed the case of Baby Doe in Bloomington, Indiana, in 1982, and the history of denying lifesaving care to infants simply because of their diagnosed disability. In 2012, Aaron was given the same level of treatment given a child who does not have a cognitive disability. While we were in the hospital in Toronto, we met many other children who were scheduled to undergo heart surgery for their ASDs or for other heart conditions such as VSDs, or atrioventricular septal defects (AVSD). None of the other children we met who were undergoing surgery had Down syndrome. In fact, AVSDs are more common among children with Down syndrome than Aaron's ASD. Consequently, we do not attribute Aaron's need for surgery to the fact that he has Down syndrome. He could have been born, like Donna, without a disabling condition, yet needing heart surgery all the same. Any child could be born with a condition like an ASD, and many pediatric heart conditions are not identifiable using currently available forms of prenatal diagnosis. Even as prenatal diagnostic technology improves, and as parental control over factors like the fetal genome is increased, children will still ultimately be able to surprise us and rebel from our control. Sometimes our children are born and need levels of care that were unanticipated in utero. Attending to these needs is part of what it means to be a parent and to have children.

The fact that Aaron was given such a high level of treatment when mere decades ago—within my lifetime—infants with Down syndrome were denied such care is a sign of moral growth among health care practitioners, and within our societies. We are more welcoming of people with cognitive disabilities into our communities than we used to be. As I argued in chapter 4, the deinstitutionalization of people with cognitive disabilities was a major step in the direction of a more welcoming attitude. With this movement, it became the norm for children to be accepted into their own families rather than to be sent away to institutions. Another area of moral growth that I have documented in this book occurs when a child with Down syndrome is born to unsuspecting parents, or when such a diagnosis is given prenatally and the parents decide to continue the pregnancy. Such parents often testify that their experiences give them a new outlook on life—and a new set of values. One goal of *Choosing Down Syndrome* is to show that there is room for further moral growth at the social level. Judging from the high rates of selective termination for Down syndrome, prospective parents still have massive reluctance to welcome children with this condition into their families. We have seen from parenting memoirs and from the experience of deinstitutionalization that, once these children are born, they are welcomed into the lives of their families. But when the decision is made prior to birth, a welcoming attitude is less common. Before birth, in fact, such an attitude is not the norm—or at least prospective parents tend to let other values and perceptions of people with Down syndrome have priority over a welcoming attitude.

Social influences about the value of people with Down syndrome and about the value of children generally play a role in these prenatal decisions. I have tried to make this point clearly in chapter 6 and elsewhere in this book. Images, ideas, and values coming from social influences shape an understanding of people with Down syndrome that has an impact on decisions about prenatal testing and about whether to continue or terminate pregnancies affected by Down syndrome. These images, ideas, and values often stigmatize people with Down syndrome and people with cognitive disabilities in general. As I show in chapter 8, one of the strongest of social influences is the economic system in which we live, and the perception that people with Down syndrome are not well suited for contributing to this system. As a parent of a child with Down syndrome, the high rates of selective termination for this condition signal a lack of social acceptance within our communities for people like my son. Ideas and values that result in the rejection of fetuses with Down syndrome come from the social sphere, the sphere (in a sense) outside of the welcoming nest of our family. Consequently, I have fears about Aaron's acceptance in this social sphere once he makes his way on his own—as he moves, more and more, among strangers.

Maybe some will argue that I am conflating different types of relationships. Prenatal decisions are about who to accept within the intimate sphere of a family. Social acceptance within a community involves a range of other types of less intimate relationships in areas like employment, health care, and housing. Our communities have made great strides in including people with cognitive disabilities in these less intimate relationships. Not only do children with Down syndrome tend to live with their families, but most people now recognize that denying public education to a child with Down syndrome is discriminatory. It is also easier for people with cognitive disabilities to have meaningful work and an income of their own—though these goals are not facilitated enough in many places. But I do not accept such a clear distinction between the intimate sphere of the family and the more formal public sphere of the community. Intimate relationships play a fundamental role in our well-being that cannot be replaced by more formal arrangements.

A friendship involves greater intimacy than a lease agreement or an employment contract, and sometimes having a friend is just as important as having an apartment or a job. Sometimes people with special needs also need help and compassion that are not required by the norms of more formal or public relationships. As a small example, imagine, for instance, a person with a cognitive disability who is lost while using public transit. For someone without special needs, it might be enough for a bus driver to refer the person to a transit map. For someone with these needs, however, it might be necessary to go above and beyond the normal requirements of a job, or of relationships between strangers, to help the person out. In such an instance, going above and beyond requires that the person providing help be motivated by an emotional or a sympathetic response requiring an investment in the other's well-being. I want Aaron to be fully accepted by his community and to have access to all the kinds of human relationships that make life go well.

For me, handing Aaron over to the surgeons is a metaphor for handing him over to society. Each involves a great degree of trust, or even faith. Aaron will gradually leave our nest. He will eventually get a job and live independently of us, his parents. He will need assistance with job training, and may need some help with daily living activities. Though his brother and sister will continue to involve him in family life, Jan and I will one day no longer be able to help and care for him. We will hand Aaron over to his siblings, his peers, his friends, and to our community. But our communities have not fully adopted a welcoming attitude toward people with Down syndrome—the attitude that I would like to see prevail when Aaron begins to make his way without us. We no longer lock people with Down syndrome up into institutions, or deny them education or health care. Parents of children with Down syndrome recognize what they add to our lives within and beyond the family. But a fully welcoming attitude would involve a greater acceptance of people with Down syndrome in intimate spheres of human relationships. To make this change, we need a greater level of acceptance and welcome within public attitudes toward people with Down syndrome. The common tendency to seek out fetuses with Down syndrome prenatally and terminate such pregnancies is a sign that this norm of acceptance does not prevail. This norm will not prevail as long as people feel an urgent need to detect and terminate pregnancies affected by Down syndrome.

Notes

Preface

1. Some people object to the terms "selective abortion" or "selective termination," which are commonly used in the bioethics literature to refer to abortion for the purpose of avoiding the birth a child with a disability. I use them in that sense only, with no desire to offend anyone.

2. Baby Doe's birth mirrored a similar case from Johns Hopkins in 1963 that also provoked extensive discussion in bioethics (Gustafson 1973).

Chapter 1

1. Two other forms of Down syndrome are caused by "translocation," in which part of the twenty-first chromosome has broken off and attached to another chromosome, and by "mosaicism," in which only some cells in the body have three copies of the twenty-first chromosome, whereas others have two.

2. Noninvasive prenatal testing is also often referred to as "cell-free fetal DNA testing" or "cffDNA testing." Throughout this book, I will use the NIPT acronym since the "cffDNA" acronym is more common in the technical literature and among professionals and specialists, whereas the NIPT acronym appears to be used more generally.

3. The accuracy of a screening or diagnostic test can be measured in a number of different ways. "Positive predictive value" is a measure of accuracy that differs from "sensitivity" and "specificity." Positive predictive value is the probability that someone with a positive test result is in fact affected by the condition for which the person is being tested (Nuffield Council on Bioethics 2017). Unlike sensitivity and specificity, the positive predictive value of a test is affected by the prevalence of the condition in the population being tested. Positive predictive value for noninvasive prenatal testing in cases of Down syndrome increases when the test is being used in a population for which there is a higher likelihood that the fetus has Down syndrome—for example, when NIPT is used as a second-tier screening test, as opposed

to a first-tier test. In the clinical setting, the positive predictive value of a test tends to provide more relevant information about the accuracy of the test for patients and clinicians than the sensitivity and specificity of a test.

4. When referring to "maternal serum screening," I mean the triple-screen test used to detect alpha-fetoprotein (AFP), human chorionic gonadotropin (hCG), and estriol in a pregnant woman's blood. Abnormal levels of these substances indicate the possibility that the fetus has spina bifida or a genetic condition such as Down syndrome. The results of maternal serum screening combined with factors such as the pregnant woman's ethnicity, weight, and gestational age of the fetus provide a probability that the fetus has one of these conditions.

Chapter 2

1. In this book I use the terms such as "well-being," "quality of life," "welfare" as interchangeable. However, one challenging concept is the term "flourishing." As will be seen later in this chapter, Jonathan Glover (2006) holds that people with disabilities tend not to flourish. He takes "flourishing" to have an Aristotelian and objective aspect, which does not cohere well with my position that subjective self-assessments of well-being are largely authoritative. Although I acknowledge Glover's interpretation of "flourishing," I disagree with it. I believe that people with disabilities are able to flourish, and I think that their testimonials to such flourishing should be believed.

2. According to a common taxonomy of theories of well-being, hedonistic and desire-fulfillment theories of well-being are "subjective" in nature because they hold that a person's well-being is dependent upon their subjective states or "pro-attitudes" (Parfit 1984; Scanlon 1998). Subjective theories are contrasted with "objective list" theories that provide a substantive account of what makes life go well. Nussbaum's capabilities approach is an example of an objective list theory (Nussbaum 2006). Since I endorse the authority of subjective reports about the well-being of people with Down syndrome and other disabilities, it might seem that I am committed to a subjective theory of well-being. But I am not. When researchers ask people with Down syndrome what makes their lives go well, they tend to discuss the sorts of things in objective list theories, such as friendships, family life, enjoyable and rewarding activities, rather than describing a good life in terms of hedonistic sensations or subjective desire fulfillment. Wasserman and Asch (2014, 142) argue that "subjective accounts based on pleasure or satisfaction are not necessarily better at recognizing the possibilities for doing well with a disability than objective or pluralistic accounts based on activity and achievement." I believe it is possible to defend an objective list theory of well-being, while also taking subjective reports of well-being by people with Down syndrome to be authoritative. People living with disabilities have a privileged *epistemic* position regarding whether their lives contain the elements that contribute to their well-being—elements that are best captured by objective list theories.

3. A more general theory of well-being and living with Down syndrome would have to account for certain puzzles associated with the high degree of well-being of people with this condition. For instance, if there was a treatment for the cognitive disabilities associated with Down syndrome, it might be considered beneficial or encouraging—even by many people with Down syndrome—to give them this treatment. Such a treatment could essentially eliminate the disability. But if people with the disability already have a high degree of well-being, on what grounds would administering the treatment be justified? Why provide treatment for a disability if the disability does not make the lives of those who have it go poorly? Without offering a general theory of well-being here, I see such examples as akin to other puzzles of well-being that do not involve disability. For instance, if a child has a high degree of well-being, how can changing into an adult contribute to, or be consistent with, that child's high well-being? If a child's life is going well, the view that it is good for the child to become an adult might seem odd. Similarly, Wasserman and Asch (2014) give an example of a person who is illiterate, yet still has a high degree of well-being. In such a case, on what grounds would it be beneficial for that person to learn how to read? We think literacy would be beneficial, but do not want to claim, contrary to the evidence, that illiteracy necessarily makes a person's life go badly. Hans Reinders (2014) proposes a solution to such puzzles with his "dynamic theory" of well-being, which is influenced by Nussbaum, and ultimately by Aristotle. Reinders proposes that an adequate theory of well-being should take into account the unfinished nature of our future lives and our capacity for growth: "Human beings flourish to the extent that they are enabled to develop their own capabilities" (Reinders 2014, 211). The ability to grow and change in the future should not be seen as contradictory to a person's high degree of well-being in the present. In most cases, the person's well-being is promoted by future growth and development.

Chapter 3

1. The baby might have certain physical health problems, however, such as a heart defect, that are sometimes associated with Down syndrome. These health problems can require immediate attention, so in such cases the baby's health needs would be an unavoidable marker of difference. Interestingly, authors of Down syndrome parenting memoirs often make a distinction between Down syndrome proper, as a cause of cognitive disabilities, and the baby's physical health problems. For just three examples out of many, see Soper (2009b, 83); Hodson (2007, 35); Cornish (2009, 232). They make this distinction as if to point out that any child could be born with a health problem, but that Down syndrome in itself is something different that requires its own coping strategies.

2. See also the statement about a new definition of "perfection," in Braat (2007, 228).

3. Some findings in neurobiology also support the idea that people tend to approach ethical problems from a virtue ethics standpoint (e.g., Casebeer 2003).

4. I thank an anonymous peer reviewer for pointing this out.

5. Hans Reinders (2000) also makes this point.

Chapter 4

1. Peter Singer continues to hold that parents should be allowed to kill their disabled newborns. In a 2016 panel discussion on Australian television, when asked about his position on infanticide, he said, "I haven't changed my mind." Singer explained by saying, "What I'm doing is trying to give parents a say in questions where they're the ones who are going to be forced to look after this child whether they want to or not" (as qtd. in Srinivasan 2016). Such a position could be motivated by a belief that parenting a child with a disability is overly burdensome. Alternatively, Singer could have retreated from endorsing the faulty empirical evidence about family functioning and turned to a position that simply supports parental autonomy over the life and death of their newborns. When faced with the fact that the research on family functioning from the 1970s and early 1980s is flawed, bioethicists tend not to change their position, but instead amend their position and claim it is supported by the principle of parental autonomy, regardless of what the empirical evidence shows (Ferguson, Gartner, and Lipsky 2000). However, the principle of parental autonomy by itself seems an insufficient justification without some empirical support showing that parents need the option of infanticide to avoid a horrible outcome for either the family or the infant. In the case of Down syndrome, this empirical support is lacking. So if Singer continues to endorse the infanticide option for infants with Down syndrome—he says he hasn't changed his mind—then either he continues to believe that raising a child with Down syndrome is devastating, or his position lacks sufficient justification.

2. This study has a number of odd inconsistencies and omissions. Though only 67 percent of respondents for the families of children with Down syndrome expressed satisfaction with the health of their families, the authors report that "general family health was seen as high in most families (92% for Down syndrome ...)" (Brown et al. 2006, 242). They don't explain how 92 percent of the respondents can characterize their family health as "high," while only 67 percent of the same group are satisfied with their family health. Furthermore, the authors say that the different areas of family life covered in their survey (health, finances, external support, careers, family relations, and so on) are essential for good quality family life, yet they apparently did not ask respondents to rate the quality of their family lives overall.

3. More recent research incorporates variables such as access to social services in efforts to study the well-being of families (e.g. Skotko, Levine, and Goldstein 2011b).

Chapter 5

1. Dan W. Brock defends a similar, though abbreviated, form of this argument in his article "Preventing Genetically Transmitted Disabilities While Respecting Persons with Disabilities" (2005). Brock also develops his supporting account of quality of life in his article "Quality of Life Measures in Health Care and Medical Ethics" (1993). I will also make reference to these sources.

2. In their account, discussed later in this chapter, Julian Savulescu and Guy Kahane (2009) run into the same problem. Without explaining who is harmed or wronged, these authors simply assert that their account of the moral obligation to choose a child without a disability is consistent with an "impersonal version" of procreative ethics (277).

3. Concepts of well-being and beliefs about valued human capacities vary widely from culture to culture. Medieval Christian culture would consider the capacity for piety a key to well-being, whereas an animistic culture might highly value the ability to dream. In cultures with a highly structured social hierarchy, talents that do not correspond to a person's station in life could undermine, rather than promote that person's well-being.

4. I argue in chapter 2 that self-assessments of well-being should be considered authoritative in most cases. People of differing abilities all tend to state that their lives are going well (Amundson 2005). I see little reason to believe that genetic advantages lead to higher levels of well-being, that we must compete in order to achieve higher levels of well-being, or even that well-being can be accumulated and compared.

5. This is not to say that all women will regard selective abortion as the rejection of a child, or that they should. The point here is only to illustrate the difficulty of establishing a moral obligation to selectively abort when many women, even those who are pro-choice, would view a fetus as a child at some stage of their pregnancies. This is a morally relevant feature of selective abortion decisions that *From Chance to Choice* (Buchanan et al. 2000) handles inadequately.

Chapter 6

1. Different variations of the objection raised by proponents of the disability critique attribute the moral wrong either to the prospective parents who decide to test and abort or to larger groups such as the medical profession or society at large, which encourage, support, or make possible widespread prenatal testing and abortion of fetuses with disabling conditions. Generally, however, most proponents of the disability critique are reluctant to find fault with individual prospective parents. Although I tend to be sympathetic to this view, my arguments are focused on prospective parents since it is in their power to choose to bring children with Down syndrome into the world.

2. One version of the expressivist argument may not be addressed by this objection. An expressivist argument may be directed at clinical guidelines that permit access to amniocentesis only when women have a maternal serum screen result that is greater than 1 in 250 that the fetus has a disability. In such guidelines, 1 in 250 is the cutoff because the likelihood of having a child with a disability like Down syndrome is then greater than the likelihood that amniocentesis will cause a miscarriage. Such a guideline implies that having a child with a disability is as bad as losing a wanted pregnancy. A guideline like this seems sufficiently rule governed from a semantic standpoint to convey this negative message. One line of expressivist argument would construe a noninvasive prenatal diagnostic test as more ethically justifiable than an invasive test because it would not carry this kind of negative message. I find this line of argument to be deficient because the likely increase in selective terminations occasioned by the widespread use of NIPT will overshadow that fact that the new test is not associated with this negative message.

3. Adrienne Asch and David Wasserman (2005) advance a similar argument that prenatal testing and termination are a rejection of intimacy.

4. Asch (2003) also argues vigorously that legislation like the Americans with Disabilities Act has not led to greater inclusion, even in areas within the reach of the law.

5. I thank an anonymous peer reviewer for suggesting this line of argument.

Chapter 7

1. One way of bridging the fact-value divide is by arguing, on Aristotelian grounds, that people, animals, and things have a telos, which constitutes proper functioning. Something or someone that does not function properly is therefore defective in both the factual and the normative sense since the physical world has norms "built in," according to the Aristotelian worldview as argued by Alasdair MacIntyre (1984). To the extent that our genes are seen as constituting our identity, essence, or telos, a genetic disorder would be likely to be viewed as both a factual and a normative defect.

2. Although some experts believe there may be a relationship between having an XYY trisomy and learning difficulties or behavior problems, other experts think this "relationship" might simply reflect the fact that such issues are themselves reasons for getting a karyotype, without which there would be no reason to suspect an XYY trisomy in the first place (Goobie and Prasad 2012).

3. Because I am less confident about the social contribution to diseases or health problems, I will just focus on cognitive differences here.

Chapter 8

1. Though it sounds like a contradiction in terms, "Progressive Conservative" is indeed the name of the party.

2. Rosemarie Garland-Thomson's insights about how disability is viewed in the context of modernity are also relevant here. Garland-Thomson (2012, 352) argues that "disability and illness frustrate modernity's investment in controlling the future." Disabilities are perceived as sources of inefficiency and as incompatible with modern systems of economic, technological, and social control.

References

Abbeduto, L., Seltzer, M. M., Shattuck, P., Krauss, M. W., Orsmond, G., and Murphy, M. M. (2004). Psychological Well-Being and Coping in Mothers of Youths with Autism, Down Syndrome, or Fragile X Syndrome. *American Journal of Mental Retardation, 109*, 237–254.

ACOG (American College of Obstetricians and Gynecologists). (2015). ACOG Statement on cfDNA Screening and Practice Advisory. https://www.acog.org/About-ACOG/News-Room/Statements/2015/ACOG-Statement-on-cfDNA-Screening-and-Practice-Advisory/.

Adams, R. (2013). *Raising Henry: A Memoir of Motherhood, Disability, and Discovery.* New Haven: Yale University Press.

Agarwal, A., Sayres, L. C., Cho, M. K., Cook-Deegan, R., and Chandrasekharan, S. (2013). Commercial Landscape of Noninvasive Prenatal Testing in the United States. *Prenatal Diagnosis, 33*, 521–531.

Albrecht, G. L., and Devlieger, P. J. (1999). The Disability Paradox: High Quality of Life against All Odds. *Social Science & Medicine, 48* (8), 977–988.

Amundson, R. (2005). Disability, Ideology, and Quality of Life: A Bias in Biomedical Ethics. In D. Wasserman, J. Bickenbach, and R. Wachbroit (Eds.), *Quality of Life and Human Difference: Genetic Testing, Health Care, and Disability* (pp. 101–114). New York: Cambridge University Press.

Anderson, K. (2007). Where There's a Will, There's a Way. In K. L. Soper (Ed.), *Gifts: Mothers Reflect on How Children with Down Syndrome Enrich Their Lives* (pp. 103–106). Bethesda, MD: Woodbine House.

Aristotle. (1984). *Nicomachean Ethics.* In J. Barnes (Ed.), *The Complete Works of Aristotle.* Princeton: Princeton University Press.

Armstrong, A. (2009). This Walker Doesn't Match My Drapes! In K. L. Soper (Ed.), *Gifts 2: How People with Down Syndrome Enrich the World* (pp. 18–21). Bethesda, MD: Woodbine House.

Asch, A. (2000). Why I Haven't Changed My Mind about Prenatal Diagnosis: Reflections and Refinements. In E. Parens and A. Asch (Eds.), *Prenatal Testing and Disability Rights* (pp. 234–258). Washington, DC: Georgetown University Press.

Asch, A. (2003). Disability Equality and Prenatal Testing: Contradictory or Compatible? *Florida State University Law Review, 30,* 315–342.

Asch, A., and Wasserman, D. (2005). Where Is the Sin in Synecdoche? Prenatal Testing and the Parent-Child Relationship. In D. Wasserman, J. Bickenbach, and R. Wachbroit (Eds.), *Quality of Life and Human Difference: Genetic Testing, Health Care, and Disability* (pp. 172–216). New York: Cambridge University Press.

Bailey, C. J. (2009). Heroes. In K. L. Soper (Ed.), *Gifts 2: How People with Down Syndrome Enrich the World* (pp. 214–217). Bethesda, MD: Woodbine House.

Baily, M. A. (2000). Why I Had Amniocentesis. In E. Parens and A. Asch (Eds.), *Prenatal Testing and Disability Rights* (pp. 64–71). Washington, DC: Georgetown University Press.

Barnes, E. (2016). *The Minority Body: A Theory of Disability.* New York: Oxford University Press.

Bauer, P. E. (2005). The Abortion Debate No One Wants to Have. *Washington Post,* October 18. http://www.washingtonpost.com/wp-dyn/content/article/2005/10/17/AR2005101701311_pf.html.

Bazelon, E. (2010). The New Abortion Providers. *New York Times Magazine,* July 12. http://www.nytimes.com/2010/07/18/magazine/18abortion-t.html.

Beck, M. (1999). *Expecting Adam: A True Story of Birth, Rebirth, and Everyday Magic.* New York: Berkley Books.

Bennett, R. (2009). The Fallacy of the Principle of Procreative Beneficence. *Bioethics, 23* (5), 265–273.

Bernhardt, B. (2014). Genetic Counselors and the Future of Clinical Genomics. *Genome Medicine, 6* (7), 49.

Bérubé, M. (1996). *Life As We Know It: A Father, a Family, and an Exceptional Child.* New York: Pantheon Books.

Bickenbach, J. E., Felder, F., and Schmitz, B. (2014). *Disability and the Good Human Life.* New York: Cambridge University Press.

Bishop, T. F., Federman, A. D., and Keyhani, S. (2010). Physicians' Views on Defensive Medicine: A National Survey. *Archives of Internal Medicine, 170,* 1081–1083.

Bishop, T. F., and Pesko, M. (2015). Does Defensive Medicine Protect Doctors against Malpractice Claims? *British Medical Journal, 351,* h5786. doi:10.1136/bmj.h5786.

Blougouras, J. E. (2007). Notes from the Deep End. In K. L. Soper (Ed.), *Gifts: Mothers Reflect on How Children with Down Syndrome Enrich Their Lives* (pp. 61–63). Bethesda, MD: Woodbine House.

Boggian, C. (2007). Crecer Con Amour. In K. L. Soper (Ed.), *Gifts: Mothers Reflect on How Children with Down Syndrome Enrich Their Lives* (pp. 261–267). Bethesda, MD: Woodbine House.

Botkin, J. R. (2000). Line Drawing: Developing Professional Standards for Prenatal Diagnostic Services. In E. Parens and A. Asch (Eds.), *Prenatal Testing and Disability Rights* (pp. 288–307). Washington, DC: Georgetown University Press.

Braat, M. (2007). The Life That Chose Me. In K. L. Soper (Ed.), *Gifts: Mothers Reflect on How Children with Down Syndrome Enrich Their Lives* (pp. 225–229). Bethesda, MD: Woodbine House.

Branam, C. (2007). Lucky. In K. L. Soper (Ed.), *Gifts: Mothers Reflect on How Children with Down Syndrome Enrich Their Lives* (pp. 208–211). Bethesda, MD: Woodbine House.

Brink, D. O. (2014). Principles and Intuitions in Ethics: Historical and Contemporary Perspectives. *Ethics, 124*, 665–694.

Brock, D. (1993). Quality of Life Measures in Health Care and Medical Ethics. In M. C. Nussbaum and A. Sen (Eds.), *The Quality of Life* (pp. 95–132). New York: Clarendon Press.

Brock, D. (2005). Preventing Genetically Transmitted Disabilities While Respecting Persons with Disabilities. In D. Wasserman, J. Bickenbach, and R. Wachbroit (Eds.), *Quality of Life and Human Difference: Genetic Testing, Health Care, and Disability* (pp. 67–100). New York: Cambridge University Press.

Brody, H. (1994). My Story Is Broken: Can You Help Me Fix It? Medical Ethics and the Joint Construction of Narrative. *Literature and Medicine, 13* (1), 79–92.

Brown, I. (2009). *The Boy in the Moon: A Father's Search for His Disabled Son*. Toronto: Vintage Canada.

Brown, R. I., MacAdam-Crisp, J., Wang, M., and Iarocci, G. (2006). Family Quality of Life When There Is a Child with a Developmental Disability. *Journal of Policy and Practice in Intellectual Disabilities, 3* (4), 238–245.

Bryant, L. D., Green, J., and Hewison, J. (2010). The Role of Attitudes Towards the Targets of Behaviour in Predicting and Informing Prenatal Testing Choices. *Psychology and Health, 25* (10), 1175–1194.

Bubeck, D. (2004). Sex Selection: The Feminist Response. In J. Burley and J. Harris (Eds.), *A Companion to Genethics* (pp. 216–228). Malden, MA: Blackwell.

Buchanan, A., Brock, D. W., Daniels, N., and Wikler, D. (2000). *From Chance to Choice: Genetics and Justic*. New York: Cambridge University Press.

Buchanan, A., Sachs, A., Toler, T., and Tsipis, J. (2014). NIPT: Current Utilization and Implications for the Future of Prenatal Genetic Counseling. *Prenatal Diagnosis, 34* (9), 850–857.

Cadwallader, J. R. (2013). Figuring (Wounded) Detachment: Rape, Embodiment and Therapeutic Forgetting. *Somatechnics, 3* (2), 250–269.

Canadian Down Syndrome Society. (2011). Your Child with Down Syndrome. Pamphlet available from Canadian Down Syndrome Society, http://cdss.ca/.

Canadian Down Syndrome Society. (2015). Past Heroes. http://cdss.ca/be-inspired/awards/past-heroes/.

Caplan, A. L. (2015). Chloe's Law: A Powerful Legislative Movement Challenging a Core Ethical Norm of Genetic Testing. *PLoS Biology, 13* (8), e1002219.

Carlson, L. (2010). *The Faces of Intellectual Disability: Philosophical Reflections*. Indianapolis: Indiana University Press.

Carroll, J. (2009). Genetic Testing: Counselors Desperately Needed. *Biotechnology Healthcare, 6* (2), 14–22.

Casebeer, W. D. (2003). Moral Cognition and Its Neural Constituents. *Nature Reviews. Neuroscience, 4* (10), 840–846.

Castellano-Farrell, D. (2007). Brothers. In K. L. Soper (Ed.), *Gifts: Mothers Reflect on How Children with Down Syndrome Enrich Their Lives* (pp. 236–239). Bethesda, MD: Woodbine House.

CBC News. (2014). Tim Hudak Would Cut 100,000 Public Sector Jobs if Tories Win Ontario Election. *CBC News*. http://www.cbc.ca/news/canada/toronto/ontario-votes-2014/tim-hudak-would-cut-100-000-public-sector-jobs-if-tories-win-ontario-election-1.2637436.

Chambers, T. (2010). Literature. In J. Sugarman and D. P. Sulmasy (Eds.), *Methods in Medical Ethics* (2nd ed., pp. 159–174). Washington, DC: Georgetown University Press.

Chandler, J. A., Mogyoros, A., Rubio, T. M., and Racine, E. (2013). Another Look at the Legal and Ethical Consequences of Pharmacological Memory Dampening: The Case of Sexual Assault. *Journal of Law, Medicine & Ethics, 41* (4), 859–871.

Chen, S. C., and Wasserman, D. T. (2017). A Framework for Unrestricted Prenatal Whole-Genome Sequencing: Respecting and Enhancing the Autonomy of Prospective Parents. *American Journal of Bioethics, 17* (1), 3–18.

Choi, H., Van Riper, M., and Thoyre, S. (2012). Decision Making Following a Prenatal Diagnosis of Down Syndrome: An Integrative Review. *Journal of Midwifery & Women's Health, 57*, 156–164.

Coles, S., and Scior, K. (2012). Public Attitudes towards People with Intellectual Disabilities: A Qualitative Comparison of White British and South Asian People. *Journal of Applied Research in Intellectual Disabilities, 25*, 177–188.

Cornish, M. (2009). Journey to Hope. In K. L. Soper (Ed.), *Gifts 2: How People with Down Syndrome Enrich the World* (pp. 232–239). Bethesda, MD: Woodbine House.

Couser, G. T. (2009). *Signifying Bodies: Disability in Contemporary Life Writing*. Ann Arbor: University of Michigan Press.

Crawford, B. (2007). The Buddha Bean. In K. L. Soper (Ed.), *Gifts: Mothers Reflect on How Children with Down Syndrome Enrich Their Lives* (pp. 127–129). Bethesda, MD: Woodbine House.

Culp, A. (2007). Light the Way. In K. L. Soper (Ed.), *Gifts: Mothers Reflect on How Children with Down Syndrome Enrich Their Lives* (pp. 118–122). Bethesda, MD: Woodbine House.

Cunningham, C. [C.] (1982). *Down's Syndrome: An Introduction for Parents*. London: Souvenir Press.

Cunningham, C. C. (1996). Families of Children with Down Syndrome. *Down's Syndrome: Research and Practice, 4* (3), 87–95.

Cuskelly, M., Chant, D., and Hayes, A. (1998). Behavioral Problems in the Siblings of Children with Down Syndrome: Associations with Family Responsibilities and Parental Stress. *International Journal of Disability Development and Education, 45* (3), 295–311.

Cuskelly, M., and Gunn, P. (2006). Adjustment of Children Who Have a Sibling with Down Syndrome: Perspectives of Mothers, Fathers, and Children. *Journal of Intellectual Disability Research, 50* (12), 917–925.

Cuskelly, M., Hauser-Cram, P., and Van Riper, M. (2008). Families of Children with Down Syndrome: What We Know and What We Need to Know. *Down's Syndrome: Research and Practice*. doi:10.3104/reviews.2079.

Daley, B. (2014). Oversold Prenatal Tests Spur Some to Choose Abortions. *Boston Globe*, December 14. https://www.bostonglobe.com/metro/2014/12/14/oversold-and-unregulated-flawed-prenatal-tests-leading-abortions-healthy-fetuses/aKFAOCP5N0Kr8S1HirL7EN/story.html.

Daniels, N. (1996). *Justice and Justification: Reflective Equilibrium in Theory and Practice*. New York: Cambridge University Press.

Daniels, N. (2012). Reasonable Disagreement about Identified vs. Statistical Victims. *Hastings Center Report, 42* (1), 35–45.

Dar, P., Curnow K. J., Gross, S. J, Hall, M. P., Stosic, M., Demko, Z., et al. (2014). Clinical Experience with Large Scale Single-Nucleotide Polymorphism–Based Noninvasive Prenatal Aneuploidy Testing. *American Journal of Obstetrics and Gynecology, 211* (5), 527.e1–527.e17.

DeJong, A., and DeWert, G. M. W. R. (2015). Prenatal Screening: An Ethical Agenda for the Near Future. *Bioethics, 29* (1), 46–55.

de Melo-Martín, I. (2004). On Our Obligation to Select the Best Children: A Reply to Savulescu. *Bioethics, 18* (1), 72–83.

Donley, G., Hull, S. C., and Berkman, B. E. (2012). Prenatal Whole Genome Sequencing: Just Because We Can, Should We? *Hastings Center Report, 42* (4), 28–40.

Dreyfus, H., and Kelly, S. D. (2011). *All Things Shining: Reading the Western Classics to Find Meaning in a Secular Age*. New York: Free Press.

Duis, S. S., Summers, M., and Summers, C. R. (1997). Parent versus Child Stress in Diverse Family Types: An Ecological Approach. *Topics in Early Childhood Special Education, 17* (1), 53–73.

Dumas, J. E., Wolf, L. C., Fisman, S. N., and Culligan, A. (1991). Parenting Stress, Child Behavior Problems, and Dysphoria in Parents of Children with Autism, Down Syndrome, Behavior Disorders, and Normal Development. *Exceptionality: A Special Education Journal, 2* (2), 97–110.

Ehrich, M., Deciu, C., Zwiefelhofer, T., Tynan, J. A., Cagasan, L., Tim, R., et al. (2011). Noninvasive Detection of Fetal Trisomy 21 by Sequencing of DNA in Maternal Blood: A Study in a Clinical Setting. *American Journal of Obstetrics and Gynecology, 204*, 205.e1–205.e11.

Elster, J. (2011). Procreative Beneficence—Cui Bono? *Bioethics, 25* (9), 482–488.

Enea-Drapeau, C., Carlier, M., and Huguet, P. (2012). Tracing Subtle Stereotypes of Children with Trisomy21: From Facial-Feature-Based to Implicit Stereotyping. *PLoS One, 7* (4), e34369.

Estreich, G. (2011). *The Shape of the Eye: Down Syndrome, Family, and the Stories We Inherit*. Dallas: Southern Methodist University Press.

Farrelly, C. (2007). Virtue Ethics and Prenatal Genetic Enhancement. *Studies in Ethics, Law, and Techology, 1*, (1), article 4.

Ferguson, P. M., Gartner, A., and Lipsky, D. K. (2000). The Experience of Disability in Families: A Synthesis of Research and Parent Narratives. In E. Parens and A. Asch

(Eds.), *Prenatal Testing and Disability Rights* (pp. 72–94). Washington, DC: Georgetown University Press.

Foraker, B. (2009). Learning to Fly. In K. L. Soper (Ed.), *Gifts 2: How People with Down Syndrome Enrich the World* (pp. 36–38). Bethesda, MD: Woodbine House.

Forman, V. (2009). *This Lovely Life: A Memoir of Premature Motherhood*. New York: Houghton Mifflin Harcourt.

Francis. (2015). Encyclical Letter *Laudato Si'* of the Holy Father Francis on Care for Our Common Home. http://w2.vatican.va/content/francesco/en/encyclicals/documents/papa-francesco_20150524_enciclica-laudato-si.html.

Frank, A. W. (1995). *The Wounded Storyteller: Body, Illness, and Ethics*. Chicago: University of Chicago Press.

Fraser, N., and Gordon, L. (1994). Civil Citizenship against Social Citizenship? In B. van Steenbergen (Ed.), *The Condition of Citizenship* (pp. 90–107). New York: Sage.

Gallimore, R., Weisner, T. S., Kaufman, S. Z., and Bernheimer, L. P. (1989). The Social Construction of Ecocultural Niches: Family Accommodation of Developmentally Delayed Children. *American Journal of Mental Retardation, 94* (3), 216–230.

Garland-Thomson, R. (2012). The Case for Conserving Disability. *Bioethical Inquiry, 9*, 339–355.

Genetics Education Canada. (2017). Non-Invasive Prenatal Testing. http://genetics-education.ca/educational-resources/gec-ko-on-the-run/non-invasive-prenatal-testing/#ON.

GenomeWeb News. (2015). Sequenom Reports Preliminary Q4 Revenue of $37M. *GenomeWeb*. https://www.genomeweb.com/business-news/sequenom-reports-preliminary-q4-revenue-37m.

Gilligan, C. (1982). *In a Different Voice*. Cambridge, MA: Harvard University Press.

Ginsburg, F., and Rapp, R. (2010). Enabling Disability: Rewriting Kinship, Reimagining Citizenship. In L. J. Davis (Ed.), *The Disability Studies Reader* (3rd ed., pp. 237–253). New York: Routledge.

Glannon, W. (2006). Psychopharmacology and Memory. *Journal of Medical Ethics, 32*, 74–78.

Global Down Syndrome Foundation. (2015). Facts and FAQ about Down Syndrome. http://www.globaldownsyndrome.org/about-down-syndrome/facts-about-down-syndrome/.

Glover, J. (2006). *Choosing Children: Genes, Disability, and Design*. New York: Oxford University Press.

Goering, S. (2008). "You Say You're Happy, but …": Contested Quality of Life Judgments in Bioethics and Disability Studies. *Bioethical Inquiry, 5,* 125–135.

Gonzales, J. (2007). The Blessing of an Imperfect Life. In K. L. Soper (Ed.), *Gifts: Mothers Reflect on How Children with Down Syndrome Enrich Their Lives* (pp. 248–252). Bethesda, MD: Woodbine House.

Goobie, S., and Prasad, C. (2012). XYY Syndrome. *eLS: Citable Reviews in the Life Sciences.* http://www.els.net/WileyCDA/ElsArticle/refId-a0005157.html.

Gramsci, A. (1971). *Selections from the Prison Notebooks.* New York: International.

Grant, R., and Flint, K. (2007). Prenatal Screening for Fetal Aneuploidy: A Commentary by the Canadian Down Syndrome Society. *Journal of Obstetrics and Gynaecology Canada, 29* (7), 580–582.

Groneberg, J. G. (2007). First Words. In K. L. Soper (Ed.), *Gifts: Mothers Reflect on How Children with Down Syndrome Enrich Their Lives* (pp. 288–292). Bethesda, MD: Woodbine House.

Groneberg, J. G. (2008). *Road Map to Holland: How I Found My Way Through My Son's First Two Years with Down Syndrome.* New York: New American Library.

Guillaume, C. (2007). Green Onion. In K. L. Soper (Ed.), *Gifts: Mothers Reflect on How Children with Down Syndrome Enrich Their Lives* (pp. 149–151). Bethesda, MD: Woodbine House.

Gustafson, J. M. (1973). Mongolism, Parental Desires, and the Right to Life. *Perspectives in Biology and Medicine, 16* (4), 529–557.

Hampton, K. (2012). *Bloom: Finding Beauty in the Unexpected.* New York: William Morrow.

Hannam, C. (1980). *Parents and Mentally Handicapped Children.* New York: Penguin Books.

Hein, S., Grumm, M., and Fingerle, M. (2011). Is Contact with People with Disabilities a Guarantee for Positive Implicit and Explicit Attitudes? *European Journal of Special Needs Education, 26* (4), 509–522.

Henry, M., Fishman, J. R., and Youngner, S. J. (2007). Propranolol and the Prevention of Post-Traumatic Stress Disorder: Is it Wrong to Erase the "Sting" of Bad Memories? *American Journal of Bioethics, 7* (9), 12–20.

Herissone-Kelly, P. (2006). Procreative Beneficence and the Prospective Parent. *Journal of Medical Ethics, 32,* 166–169.

Hodson, T. (2007). Different. In K. L. Soper (Ed.), *Gifts: Mothers Reflect on How Children with Down Syndrome Enrich Their Lives* (pp. 34–37). Bethesda, MD: Woodbine House.

Holland, S. (2011). The Virtue Ethics Approach to Bioethics. *Bioethics, 25* (4), 192–201.

Holm, S. (2008). The Expressivist Objection to Prenatal Diagnosis: Can it Be Laid to Rest? *Journal of Medical Ethics, 34*, 24–25.

Huffman, J. S. (2007). It's Better Than Good. In K. L. Soper (Ed.), *Gifts: Mothers Reflect on How Children with Down Syndrome Enrich Their Lives* (pp. 38–43). Bethesda, MD: Woodbine House.

Iannone, N. (2007). A Hopeful Future. In K. L. Soper (Ed.), *Gifts: Mothers Reflect on How Children with Down Syndrome Enrich Their Lives* (pp. 130–134). Bethesda, MD: Woodbine House.

Iles, R. K., Shahpari, M. E., Cuckle, H., and Butler, S. A. (2015). Direct and Rapid Mass Spectral Fingerprinting of Maternal Urine for the Detection of Down Syndrome Pregnancy. *Clinical Proteomics, 12* (9). doi:10.1186/s12014-015-9082-9.

International Mosaic Down Syndrome Association. (2011). MDS FAQs. http://www.imdsa.org/mdsfacts.

James, W. (1890). *Principles of Psychology*. New York: Henry Holt.

Jennings, B. (2000). Technology and the Genetic Imaginary: Prenatal Testing and the construction of Disability. In E. Parens and A. Asch (Eds.), *Prenatal Testing and Disability Rights* (pp. 124–144). Washington, DC: Georgetown University Press.

Juth, N., and Munthe, C. (2012). *The Ethics of Screening in Health Care and Medicine: Serving Society or Serving the Patient?*. New York: Springer.

Kahneman, D. (2011). *Thinking, Fast and Slow*. New York: Farrar, Strauss, and Giroux.

Kahneman, D., and Deaton, A. (2010). High Income Improves Evaluation of Life but not Emotional Well-Being. *Proceedings of the National Academy of Sciences of the United States of America, 107* (38), 16489–16493.

Kaminsky, L., and Dewey, D. (2002). Psychosocial Adjustment in Siblings of Children with Autism. *Journal of Child Psychology and Psychiatry, and Allied Disciplines, 43* (2), 225–232.

Kaposy, C. (2010a). Improving Abortion Access in Canada. *Health Care Analysis, 18* (1), 17–34.

Kaposy, C. (2010b). Proof and Persuasion in the Philosophical Debate about Abortion. *Philosophy & Rhetoric, 43* (2), 139–162.

Kaposy, C. (2012). Two Stalemates in the Philosophical Debate about Abortion and Why They Cannot Be Resolved Using Analogical Arguments. *Bioethics, 26* (2), 84–92.

Kaposy, C. (2013). A Disability Critique of the New Prenatal Test for Down Syndrome. *Kennedy Institute of Ethics Journal, 23* (4), 299–324.

Karow, J. (2016). Noninvasive Prenatal Testing Diversifies in 2015, Migrates Toward Average-Risk Market. *GenomeWeb*. https://www.genomeweb.com/molecular-diagnostics/noninvasive-prenatal-testing-diversifies-2015-migrates-toward-average-risk

Kent, D. (2000). Somewhere a Mockingbird. In E. Parens and A. Asch (Eds.), *Prenatal Testing and Disability Rights* (pp. 57–63). Washington, DC: Georgetown University Press.

Khader, S. J. (2011). *Adaptive Preferences and Women's Empowerment*. New York: Oxford University Press.

King, G. A., Zwaigenbaum, L., King, S., Baxter, D., Rosenbaum, P., and Bates, A. (2006). A Qualitative Investigation of Changes in the Belief Systems of Families of Children with Autism or Down Syndrome. *Child: Care, Health and Development, 32* (3), 353–369.

Kittay, E. F. (1999). *Love's Labor: Essays on Women, Equality, and Dependency*. New York: Routledge.

Kittay, E. F. (2005). Equality, Dignity, and Disability. In M. A. Lyons and F. Waldron (Eds.), *Perspectives on Equality: The Second Seamus Heaney Lectures* (pp. 95–122). Dublin: Liffey Press.

Kittay, E. F., and Kittay, L. (2000). On the Expressivity and Ethics of Selective Abortion for Disability: Conversations with My Son. In E. Parens and A. Asch (Eds.), *Prenatal Testing and Disability Rights* (pp. 165–195). Washington, DC: Georgetown University Press.

Kleege, G. (1999). *Sight Unseen*. New Haven: Yale University Press.

Kobelka, C., Mattman, A., and Langlois, S. (2009). An Evaluation of the Decision-Making Process Regarding Amniocentesis Following a Screen-Positive Maternal Serum Screen Result. *Prenatal Diagnosis, 29*, 514–519.

Kolata, G. (2014). Ethical Questions Arise As Genetic Testing of Embryos Increases. *New York Times*, February 3. http://www.nytimes.com/2014/02/04/health/ethics-questions-arise-as-genetic-testing-of-embryos-increases.html.

Kopp, B. (2009). Do You See What I See? In K. L. Soper (Ed.), *Gifts 2: How People with Down Syndrome Enrich the World* (pp. 114–118). Bethesda, MD: Woodbine House.

Korenromp, M. J., Page-Christiaens, G. C., van den Bout, J., Mulder, E. J., and Visser, G. H. (2007). Maternal Decision to Terminate Pregnancy after a Diagnosis of Down Syndrome. *American Journal of Obstetrics and Gynecology, 196*, 149.e1–149.e11.

Krahn, T. M. (2015). Down Syndrome and Rightful Expectations for a More Balanced Research Agenda in Canada. Twenty-eighth Annual Canadian Bioethics Society Conference, Winnipeg, May 29.

Kuhse, H., and Singer, P. (1985). *Should the Baby Live? The Problem of Handicapped Infants*. New York: Oxford University Press.

Kutner, R. (2014). Richard Dawkins: It Would Be "Immoral" Not to Abort a Fetus with Down Syndrome. *Salon.com*, August 21. http://www.salon.com/2014/08/21/richard_dawkins_it_would_be_immoral_not_to_abort_a_fetus_with_down_syndrome/.

Lantos, J. D., and Meadow, W. L. (2006). *Neonatal Bioethics: The Moral Challenges of Medical Innovation*. Baltimore: Johns Hopkins University Press.

Leach, M. W. (2009). Vanishing Beauty. In K. L. Soper (Ed.), *Gifts 2: How People with Down Syndrome Enrich the World* (pp. 93–96). Bethesda, MD: Woodbine House.

Ledford, H. (2015). CRISPR, The Disruptor. *Nature, 522*, 20–24.

LeFrançois, B., Menzies, R., and Reaume, G. (2013). *Mad Matters: A Critical Reader in Canadian Mad Studies*. Toronto: Canadian Scholars' Press.

Lester, S. (2007). Growing. In K. L. Soper (Ed.), *Gifts: Mothers Reflect on How Children with Down Syndrome Enrich Their Lives* (pp. 253–258). Bethesda, MD: Woodbine House.

Lewin, T. (2015). Ohio Bill Would Ban Abortion If Down Syndrome Is Reason. *New York Times*, August 22. http://www.nytimes.com/2015/08/23/us/ohio-bill-would-ban-abortion-if-down-syndrome-is-reason.html.

Lindeman, R. (2008). Take Down Syndrome Out of the Abortion Debate. *Canadian Medical Association Journal, 179* (10), 1088.

Lippman, A. (2004). Prenatal Genetic Testing and Screening: Constructing Needs and Reinforcing Inequities. In F. Baylis, J. Downie, B. Hoffmaster, and S. Sherwin (Eds.), *Health Care Ethics in Canada* (2nd ed., pp. 401–411). Toronto: Thomson Nelson.

Locke, J. ([1690] 1979). *An Essay Concerning Human Understanding*. New York: Oxford University Press.

Löwy, I. (2014). Prenatal Diagnosis: The Irresistible Rise of the "Visible" Fetus. *Studies in History and Philosophy of Biological and Biomedical Sciences, 47*, 290–299.

Macfarlane, J. (2015). New Blood Test Blamed As Women Choosing to Abort Babies with Down's Syndrome and Other Serious Disabilities Soars 34% in Three Years. *Mail Online*, June 13. http://www.dailymail.co.uk/news/article-3123078/New-blood-test-blamed-women-choosing-abort-babies-s-syndrome-disabilities-soars-34-three-years.html.

MacIntyre, A. (1984). *After Virtue* (2nd ed.). Notre Dame: Notre Dame University Press.

Mackenzie, C., and Stoljar, N. (2000). Introduction: Autonomy Refigured. In C. Mackenzie and N. Stoljar (Eds.), *Relational Autonomy: Feminist Perspectives on Autonomy, Agency, and the Social Self* (pp. 3–31). New York: Oxford University Press.

Maharaj, S. (2014). Tim Hudak Would Strike Major Blow to Ontario Schools. *Toronto Star*, May 13. https://www.thestar.com/opinion/commentary/2014/05/13/tim_hudak_win_would_strike_major_blow_to_ontario_schools.html.

Malek, J. (2010). Deciding against Disability: Does the Use of Reproductive Genetic Technologies Express Disvalue for People with Disabilities? *Journal of Medical Ethics, 36*, 217–221.

Mansfield, C., Hopfer, S., and Marteau, T. M. (1999). Termination Rates after Prenatal Diagnosis of Down Syndrome, Spina Bifida, Anencephaly, and Turner and Klinefelter Syndromes: A Systematic Literature Review. *Prenatal Diagnosis, 19*, 808–813.

Markens, S., Browner, C. H., and Press, N. (1999). "Because of the Risks": How U.S. Pregnant Women Account for Refusing Prenatal Screening. *Social Science & Medicine, 49*, 359–369.

Marx, K., and Engels, F. ([1845] 2004). *The German Ideology*. New York: International.

Matthews, A. (2007). Live Long, Laugh Often, Love Much. In K. L. Soper (Ed.), *Gifts: Mothers Reflect on How Children with Down Syndrome Enrich Their Lives* (pp. 278–287). Bethesda, MD: Woodbine House.

Mayo Clinic. (2017). Diseases and Conditions: Down Syndrome. http://www.mayoclinic.org/diseases-conditions/down-syndrome/home/ovc-20337339.

McDougall, R. (2005). Acting Parentally: An Argument against Sex Selection. *Journal of Medical Ethics, 31*, 601–605.

McDougall, R. (2007). Parental Virtue: A New Way of Thinking about the Morality of Reproductive Actions. *Bioethics, 21* (4), 181–190.

McGaugh, J. L. (2013). Making Lasting Memories: Remembering the Significant. *Proceedings of the National Academy of Sciences of the United States of America, 110*, 10402–10407.

McLeod, C., and Sherwin, S. (2000). Relational Autonomy, Self-Trust, and Health Care for Patients Who Are Oppressed. In C. Mackenzie and N. Stoljar (Eds.), *Relational Autonomy: Feminist Perspectives on Autonomy, Agency, and the Social Self* (pp. 259–279). New York: Oxford University Press.

McMahan, J. (2002). *The Ethics of Killing: Problems at the Margins of Life*. New York: Oxford University Press.

McMahan, J. (2013). Moral Intuition. In H. LaFollette and I. Persson (Eds.), *The Blackwell Guide to Ethical Theory* (2nd ed., pp. 103–120). Malden, MA: Wiley-Blackwell.

Meredith, S., Kaposy, C., Miller, V. J., Allyse, M., Chandrasekharan, S., and Michie, M. (2016). Impact of the Increased Adoption of Prenatal cfDNA Screening on Non-Profit Patient Advocacy Organizations in the United States. *Prenatal Diagnosis, 36* (8), 714–719.

Metzel, D. S. (2004). Historical Social Geography. In S. Noll and J. Trent (Eds.), *Mental Retardation in America: A Historical Reader* (pp. 420–444). New York: New York University Press.

Michie, M. (2016). Impact of NIPS on Abortion? It's Complicated. *Prenatal Information Research Consortium*. https://prenatalinformation.org/2016/09/16/impact-of-nips-on-abortion-its-complicated/.

Morris, S., Karlsen, S., Chung, N., Hill, M., and Chitty, L. S. (2014). Model-Based Analysis of Costs and Outcomes of Non-Invasive Prenatal Testing for Down's Syndrome Using Cell Free Fetal DNA in the UK National Health Service. *PLoS One*. doi:10.1371/journal.pone.0093559.

Mueller, V. M., Huang, T., Summers, A. M., and Winsor, S. H. M. (2005). The Influence of Risk Estimates Obtained from Maternal Serum Screening on Amniocentesis Rates. *Prenatal Diagnosis, 25*, 1253–1257.

Munson, R. (2008). *Intervention and Reflection: Basic Issues in Medical Ethics* (8th ed.). Belmont, CA: Thomson-Wadsworth.

Munthe, C. (2015). A New Ethical Landscape of Prenatal Testing: Individualizing Choice to Serve Autonomy and Promote Public Health: A Radical Proposal. *Bioethics, 29* (1), 36–45.

Murdoch, B., Ravitsky, V., Ogbogu, U., Ali-Khan, S., Bertier, G., Birko, S., et al. (2017). Non-Invasive Prenatal Testing and the Unveiling of an Impaired Translation Process. *Journal of Obstetrics and Gynaecology Canada, 39* (1), 10–17.

Murphy, T. F. (2012). *Ethics, Sexual Orientation, and Choices about Children*. Cambridge, MA: MIT Press.

National Organization on Disability. (1994). *National Organization on Disability/Louis Harris Survey of Americans with Disabilities*. New York: Louis Harris Associates.

Natoli, J. L., Ackerman, D. L., McDermott, S., and Edwards, J. G. (2012). Prenatal Diagnosis of Down Syndrome: A Systematic Review of Termination Rates (1995–2011). *Prenatal Diagnosis, 32*, 142–153.

National Commission for the Protection of Human Subjects of Biomedical and Behavioral Research. (1979). *Belmont Report*. https://www.hhs.gov/ohrp/human-subjects/guidance/belmont.html.

Nelson, H. L. (2001). *Damaged Identities, Narrative Repair*. Ithaca: Cornell University Press.

Nelson, J. L. (2000). The Meaning of the Act: Reflections on the Expressivist Force of Reproductive Decision Making and Policies. In E. Parens and A. Asch (Eds.), *Prenatal Testing and Disability Rights* (pp. 196–213). Washington, DC: Georgetown University Press.

Nelson, J. L. (2007a). Synecdoche and Stigma. *Cambridge Quarterly of Healthcare Ethics, 16*, 475–478.

Nelson, J. L. (2007b). Testing, Terminating, and Discriminating. *Cambridge Quarterly of Healthcare Ethics, 16*, 462–468.

Neonatology Today. (2011). Study Shows That New DNA Test to Identify Down Syndrome in Pregnancy Is Ready for Clinical Use. *Neonatology Today, 6* (11), 6–8.

Neufeld-Kaiser, W. A., Cheng, E. Y., and Liu, Y. J. (2015). Positive Predictive Value of Non-Invasive Prenatal Screening for Fetal Chromosome Disorders Using Cell-Free DNA in Maternal Serum: Independent Clinical Experience of a Tertiary Referral Center. *BMC Medicine, 13*, 129.

Nietzsche, F. ([1887] 1967). *On the Genealogy of Morals* (W. Kaufmann and R. J. Hollingdale, Trans.). New York: Vintage Books.

Noh, S., Dumas, J. E., Wolf, L. C., and Fisman, S. N. (1989). Delineating Sources of Stress in Parents of Exceptional Children. *Family Relations, 38* (4), 456–461.

Norton, M. E., Jacobsson, B., Swamy, G. K., Laurent, L. C., Ranzini, A. C., Brar, H., et al. (2015). Cell-Free DNA Analysis for Noninvasive Examination of Trisomy. *New England Journal of Medicine, 372* (17), 1589–1597.

Nuffield Council on Bioethics. (2017) *Non-Invasive Prenatal Testing: Ethical Issues*. https://nuffieldbioethics.org/wp-content/uploads/NIPT-ethical-issues-full-report.pdf.

Nussbaum, M. C. (2006). *Frontiers of Justice: Disability, Nationality, Species Membership*. Cambridge, MA: Belknap Press.

Nussbaum, M. C. (2011). *Creating Capabilities: The Human Development Approach*. Cambridge, MA: Belknap Press.

Nussbaum, M., and Sen, A. (1993). Introduction. In M. Nussbaum and A. Sen (Eds.), *The Quality of Life* (pp. 1–6). New York: Clarendon Press.

Ossorio, P. N. (2000). Prenatal Genetic Testing and the Courts. In E. Parens and A. Asch (Eds.), *Prenatal Testing and Disability Rights* (pp. 308–333). Washington, DC: Georgetown University Press.

Ouellette-Kuntz, H., Burge, P., Brown, H. K., and Arsenault, E. (2010). Public Attitudes Towards Individuals with Intellectual Disabilities As Measured by the Concept of Social Distance. *Journal of Applied Research in Intellectual Disabilities, 23*, 132–142.

Overall, C. (2012). *Why Have Children? The Ethical Debate*. Cambridge, MA: MIT Press.

Padua, L., Rendeli, C., Rabini, A., Girardi, E., Tonali, P., and Salvaggio, E. (2002). Health-Related Quality of Life and Disability in Young Patients with Spina Bifida. *Archives of Physical Medicine and Rehabilitation, 83*, 1384–1388.

Palomaki, G. E., Deciu, C., Kloza, E. M., Lambert-Messerlian, G. M., Haddow, J. E., Neveux, L. M., et al. (2012). DNA Sequencing of Maternal Plasma Reliably Identifies Trisomy 18 and Trisomy 13, As Well As Down Syndrome: An International Collaborative Study. *Genetics in Medicine, 14* (3), 296–305.

Palomaki, G. E., Kloza, E. M., Lambert-Messerlian, G. M., Haddow, J. E., Neveux, L. M., Ehrich, M., et al. (2011). DNA Sequencing of Maternal Plasma to Detect Down Syndrome: An International Clinical Validation Study. *Genetics in Medicine, 13* (11), 913–920.

Parens, E., and Asch, A. (2000). The Disability Rights Critique of Prenatal Genetic Testing: Reflections and Recommendations. In E. Parens and A. Asch (Eds.), *Prenatal Testing and Disability Rights* (pp. 3–43). Washington, DC: Georgetown University Press.

Parfit, D. (1984). *Reasons and Persons.* New York: Oxford University Press.

Parker, M. (2007). The Best Possible Child. *Journal of Medical Ethics, 33,* 279–283.

Parker, R. (2007). Quality of Life. In K. L. Soper (Ed.), *Gifts: Mothers Reflect on How Children with Down Syndrome Enrich Their Lives* (pp. 268–272). Bethesda, MD: Woodbine House.

Paul, D. B. (1998). *The Politics of Heredity: Essays on Eugenics, Biomedicine, and the Nature-Nurture Debate.* Albany: State University of New York Press.

PCSEPMBBR (President's Commission for the Study of Ethical Problems in Medicine and Biomedical Behavioral Research). (1983). *Screening and Counseling for Genetic Conditions.* Washington, DC: U.S. Government Printing Office.

Peele, L. (2009). Bridget's Light. In K. L. Soper (Ed.), *Gifts 2: How People with Down Syndrome Enrich the World* (pp. 86–89). Bethesda, MD: Woodbine House.

Perinatal Services BC. (2017). Non-Invasive Prenatal Testing (NIPT). http://www.perinatalservicesbc.ca/health-professionals/professional-resources/screening/prenatal-genetic/non-invasive-prenatal-testing-nipt.

Piepmeier, A. (2012). Saints, Sages, and Victims: Endorsement of and Resistance to Cultural Stereotypes in Memoirs by Parents of Children with Disabilities. *Disability Studies Quarterly, 32* (1).

Pioro, M., Mykitiuk, R., and Nisker, J. (2008). Wrongful Birth Litigation and Prenatal Screening. *Canadian Medical Association Journal, 179* (10), 1027–1030.

President's Council on Bioethics. (2003). *Beyond Therapy: Biotechnology and the Pursuit of Happiness.* Washington, DC: President's Council.

Press, N., and Browner, C. (1994). Collective Silences, Collective Fictions: How Prenatal Diagnostic Testing Became Part of Routine Prenatal Care. In K. H. Rothenberg and E. J. Thomson (Eds.), *Women and Prenatal Testing: Facing the Challenges of Genetic Technology* (pp. 201–218). Columbus: Ohio State University Press.

Price, B. (2007). When the Other Shoe Drops. In K. L. Soper (Ed.), *Gifts: Mothers Reflect on How Children with Down Syndrome Enrich Their Lives* (pp. 187–192). Bethesda, MD: Woodbine House.

Price, J. (2009). *The Woman Who Can't Forget.* New York: Free Press.

PR Newswire. (2015). Sequenom Laboratories Launches MaterniT™ GENOME. http://www.prnewswire.com/news-releases/sequenom-laboratories-launches-maternit-genome-300111863.html.

Proctor, S. N. (2012). Implicit Bias, Attributions, and Emotions in Decisions about Parents with Intellectual Disabilities by Child Protection Workers. Ph.D. diss,, Pennsylvania State University.

Rapp, R. (1998). Refusing Prenatal Diagnosis: The Meanings of Bioscience in a Multicultural World. *Science, Technology & Human Values, 23* (1), 45–70.

Rapp, R. (1999). *Testing Women, Testing the Fetus: The Social Impact of Amniocentesis in America.* New York: Routledge.

Reinders, H. (2000). *The Future of the Disabled in Liberal Society.* Notre Dame: Notre Dame University Press.

Reinders, H. (2014). Disability and Quality of Life: An Aristotelian Discussion. In J. E. Bickenbach, F. Felder, and B. Schmitz (Eds.), *Disability and the Good Human Life* (pp. 199–218). New York: Cambridge University Press.

Roberts, K. (2007). Big Sister. In K. L. Soper (Ed.), *Gifts: Mothers Reflect on How Children with Down Syndrome Enrich Their Lives* (pp. 152–155). Bethesda, MD: Woodbine House.

Robertson, J. A. (2004). Extreme Prematurity and Parental Rights after Baby Doe. *Hastings Center Report, 34* (4), 32–39.

Robey, K. L., Beckley, L., and Kirschner, M. (2006). Implicit Infantilizing Attitudes about Disability. *Journal of Developmental and Physical Disabilities, 18* (4), 441–453.

Robinson, W. (2002). The Boundaries of Identity. *Hastings Center Report, 32* (2), 45–46.

Rodrigue, J. R., Morgan, S. B., and Geffken, G. R. (1992). Psychosocial Adaptation of Fathers of Children with Autism, Down Syndrome, and Normal Development. *Journal of Autism and Developmental Disorders, 22* (2), 249–263.

Rodrigue, J. R., Morgan, S. B., and Geffken, G. R. (1993). Perceived Competence and Behavioral Adjustment of Siblings of Children with Autism. *Journal of Autism and Developmental Disorders, 23* (4), 665–674.

Rothman, D. J. (1982). Were Tuskegee and Willowbrook "Studies in Nature"? *Hastings Center Report, 12* (2), 5–7.

Rothman, D. J., and Rothman, S. M. (1983). *The Willowbrook Wars*. New York: Harper and Row.

Rothman, D. J., and Rothman, S. M. (2004). The Litigator as Reformer. In S. Noll and J. Trent (Eds.), *Mental Retardation in America: A Historical Reader* (pp. 445–465). New York: New York University Press.

Ruddick, W. (1998). Parenthood: Three Concepts and a Principle. In L. D. Houlgate (Ed.), *Family Values: Issues in Ethics, Society and the Family*. (pp. 242–251). Belmont, CA: Wadsworth.

Sackett, D. L., and Torrance, G. W. (1978). The Utility of Different Health States As Perceived by the General Public. *Journal of Chronic Diseases, 31*, 697–704.

Said, E. (1979). *Orientalism*. New York: Vintage.

Sandel, M. J. (2007). *The Case against Perfection*. Cambridge, MA: Belknap Press.

Savulescu, J. (2001). Procreative Beneficence: Why We Should Select the Best Children. *Bioethics, 15*, 413–426.

Savulescu, J. (2005). New Breeds of Humans: The Moral Obligation to Enhance. *Reproductive Biomedicine Online, 10* (Supp. 1), 36–39.

Savulescu, J., and Kahane, G. (2009). The Moral Obligation to Create Children with the Best Chance of the Best Life. *Bioethics, 23* (5), 274–290.

Sawin K. J., Brei, T. J., Buran, C. F., and Fastenau, P. S. (2002). Factors Associated with Quality of Life in Adolescents with Spina Bifida. *Journal of Holistic Nursing, 20* (3), 279–304.

Saxton, M. (2000). Why Members of the Disability Community Oppose Prenatal Diagnosis and Selective Abortion. In E. Parens and A. Asch (Eds.), *Prenatal Testing and Disability Rights* (pp. 147–164). Washington, DC: Georgetown University Press.

Saxton, M. (2010). Disability Rights and Selective Abortion. In L. J. Davis (Ed.), *The Disability Studies Reader* (3rd ed., pp. 120–132). New York: Routledge.

Scanlon, T. M. (1998). *What We Owe to Each Other*. Cambridge, MA: Harvard University Press.

Schmidt, B. (2009). Paving Roads. In K. L. Soper (Ed.), *Gifts 2: How People with Down Syndrome Enrich the World* (pp. 12–16). Bethesda, MD: Woodbine House.

Schrad, M. L. (2015). Does Down Syndrome Justify Abortion? *New York Times*, September 4. https://www.nytimes.com/2015/09/04/opinion/does-down-syndrome-justify-abortion.html.

Schroeder, S. A. (2016). Health, Disability, and Well-Being. In G. Fletcher (Ed.), *The Routledge Handbook of Philosophy of Well-Being* (pp. 221–232). New York: Routledge.

Scully, J. L., Banks, S., and Shakespeare, T. (2006). Chance, Choice and Control: Lay Debate on Prenatal Social Sex Selection. *Social Science & Medicine, 63* (1), 21–31.

Seavilleklein, V. (2009). Challenging the Rhetoric of Choice in Prenatal Screening. *Bioethics, 23* (1), 68–77.

Sequenom. (2015). MaterniT21 Plus. https://www.sequenom.com/providers/maternit21-plus/.

Shakespeare, T. (2010). The Social Model of Disability. In L. J. Davis (Ed.), *The Disability Studies Reader* (3rd ed., pp. 266–273). New York: Routledge.

Shakespeare, T. (2014). Nasty, Brutish, and Short? On the Predicament of Disability and Embodiment. In J. E. Bickenbach, F. Felder, and B. Schmitz (Eds.), *Disability and the Good Human Life* (pp. 93–112). New York: Cambridge University Press.

Sherman, R. (2007). Beginning. In K. L. Soper (Ed.), *Gifts: Mothers Reflect on How Children with Down Syndrome Enrich Their Lives* (pp. 169–173). Bethesda, MD: Woodbine House.

Sherwin, S. (1998). A Relational Approach to Autonomy in Health Care. In S. Sherwin (Ed.), *The Politics of Women's Health: Exploring Agency and Autonomy* (pp. 19–47). Philadelphia: Temple University Press.

Simon, H. (2000). Forum: A Basic Income for All. *Boston Review*. http://bostonreview.net/forum/basic-income-all/herbert-simon-ubi-and-flat-tax.

Singer, P. (1993). *Practical Ethics* (2nd ed.). New York: Cambridge University Press.

Skotko, B. G., Levine, S. P., and Goldstein, R. (2011a). Having a Brother or Sister with Down Syndrome: Perspectives from Siblings. *American Journal of Medical Genetics. Part A, 155*, 2348–2359.

Skotko, B. G., Levine, S. P., and Goldstein, R. (2011b). Having a Son or Daughter with Down Syndrome: Perspectives from Mothers and Fathers. *American Journal of Medical Genetics. Part A, 155*, 2335–2347.

Skotko, B. G., Levine, S. P., and Goldstein, R. (2011c). Self-Perceptions from People with Down Syndrome. *American Journal of Medical Genetics. Part A, 155*, 2360–2369.

Smith, D. E. (2011). The Evolution of Addiction Medicine as a Medical Specialty. *Virtual Mentor, 13* (12), 900–905. http://journalofethics.ama-assn.org/2011/12/mhst1-1112.html.

Snowdon, A. 2012. *The Sandbox Project: Strengthening Communities for Canadian Children with Disabilities*. Report for Sandbox Project's Second Annual Conference, January 19. http://sandboxproject.ca/s/2012-sandbox-conference-mh-snowdon-strengthening-communities-for-canadian-children-with-disabilities.pdf.

Solomon, A. (2012). *Far from the Tree: Parents, Children and the Search for Identity*. New York: Scribner.

Soper, K. L. (Ed.). (2007). *Gifts: Mothers Reflect on How Children with Down Syndrome Enrich Their Lives*. Bethesda, MD: Woodbine House.

Soper, K. L. (Ed.). (2009a). *Gifts 2: How People with Down Syndrome Enrich the World*. Bethesda, MD: Woodbine House.

Soper, K. L. (2009b). *The Year My Son and I Were Born: A Story of Down Syndrome Motherhood, and Self-Discovery*. Guilford, CT: GPP Life.

Sparrow, R. (2007). Procreative Beneficence, Obligation, and Eugenics. *Genomics, Society, and Policy, 3* (3), 43–59.

Sparrow, R. (2011). A Not-So-New Eugenics: Harris and Savulescu on Human Enhancement. *Hastings Center Report, 41* (1), 32–42.

Spring, L. (2007). Enjoying the Ride. In K. L. Soper (Ed.), *Gifts: Mothers Reflect on How Children with Down Syndrome Enrich Their Lives* (pp. 135–138). Bethesda, MD: Woodbine House.

Srinivasan, A. (2016). The Distinction between an Argument and Its Likely Effects. *LRB Blog*, August 16. https://www.lrb.co.uk/blog/2016/08/16/amia-srinivasan/the-distinction-between-an-argument-and-its-likely-effects/.

Steinbock, B. (2000). Disability, Prenatal Testing, and Selective Abortion. In E. Parens and A. Asch (Eds.), *Prenatal Testing and Disability Rights* (pp. 108–123). Washington, DC: Georgetown University Press.

Stoller, S. E. (2008). Why We Are Not Morally Required to Select the Best Children: A Response to Savulescu. *Bioethics, 22* (7), 364–369.

Stores, R., Stores, G., Fellows, B., and Buckley, S. (1998). Daytime Behaviour Problems and Maternal Stress in Children with Down's Syndrome, Their Siblings, and Nonintellectually Disabled and Other Intellectually Disabled Peers. *Journal of Intellectual Disability Research, 42* (3), 228–237.

Strange, H., and Chadwick, R. (2010). The Ethics of Nonmedical Sex Selection. *Health Care Analysis, 18*, 252–266.

Strong, C. (1984). The Neonatologist's Duty to Patient and Parents. *Hastings Center Report, 14* (4), 10–16.

Summers, A. M., Langlois, S., Wyatt, P., and Douglas Wilson, R. (2007). Prenatal Screening for Fetal Aneuploidy—SOGC Clinical Practice Guidelines. *Journal of Obstetrics and Gynaecology Canada, 29*, 146–161.

Szasz, T. S. (2010). *The Myth of Mental Illness: Foundations of a Theory of Personal Conduct*. New York: Harper Perennial.

Thomas, V., and Olson, D. H. (1993). Problem Families and the Circumplex Model: Observational Assessment Using the Clinical Rating Scale (CRS). *Journal of Marital and Family Therapy, 19* (2), 159–175.

Trent, J. W. (1994). *Inventing the Feeble Mind: A History of Mental Retardation in the United States*. Berkeley: University of California Press.

Ubel, P. A., Loewenstein, G., Schwarz, N., and Smith, D. (2005). Misimagining the Unimaginable: The Disability Paradox and Health Care Decision Making. *Health Psychology, 24* (4), S57–S62.

United Kingdom Department of Health. (2016). Safer Screening Test for Pregnant Women. https://www.gov.uk/government/news/safer-screening-test-for-pregnant-women.

United Kingdom National Screening Committee. (2016). *Consultation for Cell-Free DNA Testing in the First Trimester in the Fetal Anomaly Screening Programme*. London: United Kingdom National Screening Committee.

United States Catholic Bishops. (1986). Economic Justice for All: Pastoral Letter on Catholic Social Teaching and the U.S. Economy. http://www.usccb.org/upload/economic_justice_for_all.pdf.

Urbano, R. C., and Hodapp, R. M. (2007). Divorce in Families of Children with Down Syndrome: A Population-Based Study. *American Journal of Mental Retardation, 112* (4), 261–274.

van den Berg, M., Timmermans, D. R. M., Kleinveld, J. H., Garcia, E., van Vugt, J. M. G., and van der Wal, G. (2005). Accepting or Declining the Offer of Prenatal Screening for Congenital Defects: Test Uptake and Women's Reasons. *Prenatal Diagnosis, 25*, 84–90.

Van Riper, M. (2000). Family Variables Associated with Well-Being in Siblings of Children with Down Syndrome. *Journal of Family Nursing, 6* (3), 267–286.

Van Riper, M., Ryff, C., and Pridham, K. (1992). Parental and Family Well-Being in Families of Children with Down Syndrome: A Comparative Study. *Research in Nursing & Health, 15*, 227–235.

Walker, S. (2009). A Precious Gift. In K. L. Soper (Ed.), *Gifts 2: How People with Down Syndrome Enrich the World* (pp. 334–337). Bethesda, MD: Woodbine House.

Wasserman, D. (2005). The Nonidentity Problem, Disability, and the Role Morality of Prospective Parents. *Ethics, 116*, 132–152.

Wasserman, D., and Asch, A. (2007). Reply to Nelson. *Cambridge Quarterly of Healthcare Ethics, 16*, 478–482.

Wasserman, D., and Asch, A. (2014). Understanding the Relationship between Disability and Well-Being. In J. E. Bickenbach, F. Felder, and B. Schmitz (Eds.), *Disability and the Good Human Life* (pp. 139–167). New York: Cambridge University Press.

Wasserman, D., Bickenbach, J., and Wachbroit, R. (2005). Introduction. In D. Wasserman, J. Bickenbach, and R. Wachbroit (Eds.), *Quality of Life and Human Difference* (pp. 1–26). New York: Cambridge University Press.

Weir, R. (1984). *Selective Nontreatment of Handicapped Newborns*. New York: Oxford University Press.

Wendell, S. (1996). *The Rejected Body*. New York: Routledge.

Wilkinson, S. (2010). *Choosing Tomorrow's Children: The Ethics of Selective Reproduction*. New York: Oxford University Press.

Wilkinson, S. (2015). Prenatal Screening, Reproductive Choice, and Public Health. *Bioethics, 29* (1), 26–35.

Will, M. (2009). Foreword. In K. L. Soper (Ed.), *Gifts 2: How People with Down Syndrome Enrich the World* (pp. xvii–xviii). Bethesda, MD: Woodbine House.

Wilson, M. C., and Scior, K. (2014). Attitudes Towards Individuals with Disabilities As Measured by the Implicit Association Test: A Literature Review. *Research in Developmental Disabilities, 35*, 294–321.

Wilson, M. C., and Scior, K. (2015). Implicit Attitudes towards People with Intellectual Disabilities: Their Relationship with Explicit Attitudes, Social Distance, Emotions and Contact. *PLoS One*. doi:10.1371/journal.pone.0137902.

Wright, D. (2011). *Downs: The History of a Disability*. New York: Oxford University Press.

Zeid, E. (2007). Loving Emma Jayne. In K. L. Soper (Ed.), *Gifts: Mothers Reflect on How Children with Down Syndrome Enrich Their Lives* (pp. 159–165). Bethesda, MD: Woodbine House.

Index

Basic Bioethics

Arthur Caplan, editor

Books Acquired under the Editorship of Glenn McGee and Arthur Caplan

Peter A. Ubel, *Pricing Life: Why It's Time for Health Care Rationing*

Mark G. Kuczewski and Ronald Polansky, eds., *Bioethics: Ancient Themes in Contemporary Issues*

Suzanne Holland, Karen Lebacqz, and Laurie Zoloth, eds., *The Human Embryonic Stem Cell Debate: Science, Ethics, and Public Policy*

Gita Sen, Asha George, and Piroska Östlin, eds., *Engendering International Health: The Challenge of Equity*

Carolyn McLeod, *Self-Trust and Reproductive Autonomy*

Lenny Moss, *What Genes Can't Do*

Jonathan D. Moreno, ed., *In the Wake of Terror: Medicine and Morality in a Time of Crisis*

Glenn McGee, ed., *Pragmatic Bioethics, 2d edition*

Timothy F. Murphy, *Case Studies in Biomedical Research Ethics*

Mark A. Rothstein, ed., *Genetics and Life Insurance: Medical Underwriting and Social Policy*

Kenneth A. Richman, *Ethics and the Metaphysics of Medicine: Reflections on Health and Beneficence*

David Lazer, ed., *DNA and the Criminal Justice System: The Technology of Justice*

Harold W. Baillie and Timothy K. Casey, eds., *Is Human Nature Obsolete? Genetics, Bioengineering, and the Future of the Human Condition*

Robert H. Blank and Janna C. Merrick, eds., *End-of-Life Decision Making: A Cross-National Study*

Norman L. Cantor, *Making Medical Decisions for the Profoundly Mentally Disabled*

Margrit Shildrick and Roxanne Mykitiuk, eds., *Ethics of the Body: Post-Conventional Challenges*

Alfred I. Tauber, *Patient Autonomy and the Ethics of Responsibility*

David H. Brendel, *Healing Psychiatry:Bridging the Science/Humanism Divide*

Jonathan Baron, *Against Bioethics*

Michael L. Gross, *Bioethics and Armed Conflict: Moral Dilemmas of Medicine and War*

Karen F. Greif and Jon F. Merz, *Current Controversies in the Biological Sciences: Case Studies of Policy Challenges from New Technologies*

Deborah Blizzard, *Looking Within: A Sociocultural Examination of Fetoscopy*

Ronald Cole-Turner, ed., *Design and Destiny: Jewish and Christian Perspectives on Human Germline Modification*

Holly Fernandez Lynch, *Conflicts of Conscience in Health Care: An Institutional Compromise*

Mark A. Bedau and Emily C. Parke, eds., *The Ethics of Protocells: Moral and Social Implications of Creating Life in the Laboratory*

Jonathan D. Moreno and Sam Berger, eds., *Progress in Bioethics: Science, Policy, and Politics*

Eric Racine, *Pragmatic Neuroethics: Improving Understanding and Treatment of the Mind-Brain*

Martha J. Farah, ed., *Neuroethics: An Introduction with Readings*

Jeremy R. Garrett, ed., *The Ethics of Animal Research: Exploring the Controversy*

Books Acquired under the Editorship of Arthur Caplan

Sheila Jasanoff, ed., *Reframing Rights: Bioconstitutionalism in the Genetic Age*

Christine Overall, *Why Have Children? The Ethical Debate*

Yechiel Michael Barilan, *Human Dignity, Human Rights, and Responsibility: The New Language of Global Bioethics and Bio-Law*

Tom Koch, *Thieves of Virtue: When Bioethics Stole Medicine*

Timothy F. Murphy, *Ethics, Sexual Orientation, and Choices about Children*

Daniel Callahan, *In Search of the Good: A Life in Bioethics*

Robert Blank, *Intervention in the Brain: Politics, Policy, and Ethics*

Gregory E. Kaebnick and Thomas H. Murray, eds., *Synthetic Biology and Morality: Artificial Life and the Bounds of Nature*

Dominic A. Sisti, Arthur L. Caplan, and Hila Rimon-Greenspan, eds., *Applied Ethics in Mental Healthcare: An Interdisciplinary Reader*

Barbara K. Redman, *Research Misconduct Policy in Biomedicine: Beyond the Bad-Apple Approach*

Russell Blackford, *Humanity Enhanced: Genetic Choice and the Challenge for Liberal Democracies*

Nicholas Agar, *Truly Human Enhancement: A Philosophical Defense of Limits*

Bruno Perreau, *The Politics of Adoption: Gender and the Making of French Citizenship*

Carl Schneider, *The Censor's Hand: The Misregulation of Human-Subject Research*

Lydia S. Dugdale, ed., *Dying in the Twenty-First Century: Towards a New Ethical Framework for the Art of Dying Well*

John D. Lantos and Diane S. Lauderdale, *Preterm Babies, Fetal Patients, and Childbearing Choices*

Harris Wiseman, *The Myth of the Moral Brain*

Arthur L. Caplan and Jason Schwartz, eds., *Vaccine Ethics and Policy: An Introduction with Readings*

Tom Koch, *Ethics in Everyday Places: Mapping Moral Stress, Distress, and Injury*

Nicole Piemonte, *Afflicted: How Vulnerability Can Heal Medical Education and Practice*

Chris Kaposy, *Choosing Down Syndrome: Ethics and New Prenatal Testing Technologies*

Printed in the United States
by Baker & Taylor Publisher Services